FOR THE EARTH TO LIVE

The Case for Ecosocialism

by Allan Todd

Foreword by Professor Julia Steinberger

*For all capitalism's victims – past and present –
and for 'the unborn of the future.'*

For the Earth to Live: The Case for Ecosocialism
By Allan Todd
Foreword by Professor Julia Steinberger

Published 2025 by Resistance Books (London).
Copyright © Allan Todd
Design and typeset: Adam Di Chiara
Cover image: www.pexels.com/@mdsnmdsnmdsn/

ISBN 978-1-872242-25-5 (pbk)

Resistance Books (London)
info@resistancebooks.org –www.resistancebooks.org

CONTENTS

ABOUT THE AUTHOR

Allan Todd is an ecosocialist/environmental and anti-fascist activist. He is a member of Anti-Capitalist Resistance and Extinction Rebellion North Lakes (Cumbria), and is the author of *Revolutions 1789-1917* (CUP), *Trotsky: The Passionate Revolutionary* (Pen & Sword), *Ecosocialism Not Extinction* (Resistance Books), and *Che Guevara: The Romantic Revolutionary* (Pen & Sword).

ACKNOWLEDGEMENTS

Much of the thinking in this book stretches back over several years – following, much too slowly, a fuse that was first lit by reading some extracts from Rachel Carson's *Silent Spring* back in 1963.

Over the intervening sixty years, the evolution of my ideas concerning politics and the natural world owes a great deal to conversations and relationships with a large number of people – beginning with former History students such as Nicky and Kate Barrell, and Patrick Fitzgerald. I must also give special thanks to Brian Heron who, whilst I was doing history research at Lancaster University in the early 1970s, introduced me to the writings of Ernest Mandel who, for many years, was the principal theoretician of the Fourth International.

I thank, too, several long-time good friends for kindly reading, and offering constructive criticism on, early drafts of various chapters – especially Charlotte Christensen, Cynthia Rico, Hazel Graham, and Penelope Read: each of whom also recommended several useful texts which otherwise would not have come to my attention. Special thanks are also due to Jacqui Petrie, Chrissie Parfitt, Ken and Di Barrell, and Ian Rycroft, for commenting on parts of the book at various stages.

Also, formative have been discussions with friends and members of XR North Lakes; in particular, mention must be made of the intellectual stimulus provided by Anne-Marie Williams and Fiona Prior, who not only gave much of their precious time to share ideas and advice, but also suggested relevant articles and books that had passed me by. Mention should also be made of Terry Sloan, Jo Alberti and Joe Human – members of Sustainable Keswick – discussions with whom always proved useful.

I have also had valuable input over the past few years from various comrades in Anti-Capitalist Resistance, Left Unity, and Transform. Particularly helpful have been Alan Thornett, Joseph Healey, Fred Leplat, Terry Conway, Susan Pashkoff, Simon Hannah, Chris Jones, Ian Parker, Dave Kellaway, Len Arthur, Jim Hollinshead, Felicity Dowling, Bob Whitehead, Michael Tucker,

Philip Ward, and the late Neil Faulkner – all of whom have helped shape my ideas, and/or have pointed me in the direction of useful sources.

I must also thank the wonderful Ruby Turner, for giving her permission to use brief extracts from two of her songs!! Thanks are also due the team at Resistance Books and the designer Adam Di Chiara. To all those I've forgotten to mention: my sincere apologies!

There is very little – if any! – original thinking in this book: instead, what's been attempted is an assimilation, incorporation and re-ordering of the main findings and contributions of a range of leading experts on the various subject matters covered in these pages.

Unfortunately, it's transpired I've neither time, energy nor space to cover, at sufficient length, all the aspects I would have liked to have addressed in this book – for those omissions, my sincerest apologies. Any errors are, of course, down to me alone – or possibly also to my cat, Hedgie, who frequently expressed her growing concern over capitalism's destruction of the natural world, by wandering across the keyboard, often adding/deleting text in the process. Her 'comments' about climate-crisis deniers like Trump and Farage were, fortunately, indecipherable!

I would also like to thank our daughters, Megan and Vanessa – and our grandchildren, Alexander and Emilia – for continuing to provide what Coleridge described as 'a sheltering tree.' I hope that the ecosocialist and climate movements will soon provide them – and future generations – with an ecologically-sustainable and socially-just planet that acts as *their* 'sheltering tree', and which allows them, and all other Earthlings, to thrive in the coming decades.

Finally, as ever, I owe a huge debt of gratitude to Cyn, my life companion, who was 'green' before it was as widespread a political philosophy as it is now. It is to my undying shame that, despite first becoming aware – through Rachel Carson's *Silent Spring*, and my own observations of the growing negative impacts of pesticides and modern agriculture on wildlife in 1960s' rural south Norfolk – I largely ignored what were Cyn's early ecological warnings, and instead devoted most of my energies to various other political campaigns.

Whilst opposition to the Vietnam War, apartheid in South Africa, nuclear weapons and the far right – to mention just a few of those campaigns – was undoubtedly important, I tended to sideline crucial 'green' issues. So, I hope this book helps increase the flames of resistance and revolution against capitalism's forces of destruction, and thus in some way makes up for my earlier failings. As Cyn often said to me before I became an *ecosocialist*: 'But, Allan – you can't have socialism on a dead planet.'

Allan Todd
20 November 2024
Keswick

FOREWORD

Julia Steinberger
*Professor of Ecological Economics,
University of Lausanne (Switzerland)*

The book in your hands is not like other books about the climate or the environment. *For the Earth to Live: The Case for Ecosocialism* is a wholly original creation, from a wholly original author. In this Foreword, I will try to explain why I think this book is so important, and I will try to do so quickly, so you can get to the book itself as quickly as possible (which you could do immediately by turning a couple of pages!).

This book, much like its scholar-activist author Allan Todd, stands out by doing several revolutionary things at once.

First, this book is unapologetic.
This book is quite simply about, and for, ecosocialism. In a time when people are often afraid to express themselves directly, this book openly and proudly proclaims its analysis and position, making the case for a political direction that is unapologetically, uncompromisingly for ecology AND for socialism, for people and for the living Earth. Reading *For the Earth to Live* is a bracing change from the vacuous distractions, stuffy compromising and timid understatements that characterize so much of our media, political, and, yes, even scientific landscapes. Quoting the Italian communist Antonio Gramsci from start to finish, it seeks to combine 'Pessimism of the intellect, optimism of the will.' As such, it's a continuous wake-up call (Gramsci's pessimism of the intellect) and a call to action (his optimism of the will).

Secondly, this book is well-informed.
This book is a treasure-trove of well-argued and well-cited political, historical and scientific analysis. It fully discusses the scientific evidence of the climate, biodiversity and health threats we face, bringing these into the context of our political and economic systems, and

finally providing us with a revolutionary ecosocialist politics capable of facing them.

The book is divided into roughly three phases, each building upon the others: the ecological threats we face, the political and economic threats we face, and the necessity for revolutionary ecosocial politics.

To start with, Allan Todd highlights the dire warnings of prominent climate, ecological and pandemic-health science reports, many of which were already being realized as he was writing. The first three chapters, on climate, biodiversity and pandemic threats, each paint a picture of interconnected crises in our environment, and each showcase current political failures to act upon them (unless causing them to accelerate is counted as effective political action, of course). These chapters, like the rest of the book, are thoroughly referenced, making them an excellent basis for readers to go deeper and establish their own understanding.

Following upon the ecological threats, Chapters 4 and 5 delve into the threats posed by our current political systems, covering the neoliberal takeover of our democracies, and the accompanying rise of fascism. To explain how this state of affairs came about, Chapters 6 and 7 go into the colonial past and neoliberal present of capitalism, and Chapter 8 identifies the actions of 'killer corporations' as the ultimate drivers of this deadly capitalism.

Based upon this full environmental, scientific, political, economic and historical picture of our present predicament, the book turns to who we are, and what we can do. Chapter 9 illuminates a vision of human nature fully at odds with our current societies, but very much in harmony with our environment, while Chapter 10 paints the picture of a possible better world, and Chapter 11 calls for ecosocial politics.

In these final chapters, Allan Todd draws from his wide knowledge of anarchist, socialist, communist, indigenous and anti-colonial writers to make the case for a revolutionary ecosocial politics capable of bringing them together.

Thirdly, this book is compelling.
Here I have a confession to make: I have long called myself an ecosocialist, simply because I see my life purpose as attempting to contribute to both social emancipation and well-being within planetary boundaries. In fact, this is the whole focus of my research: trying

to quantify the resources required for universal human well-being. I also had another, more superficial, reason to introduce myself as ecosocialist: in our neoliberalism-dominated times, ecosocialism still carries a pleasing shock value. It has the uncanny ability to attract the people I want to talk and work with, and displease those I would rather avoid. Indeed, it shares this characteristic with another idea that defines my work, and features prominently in *For the Earth to Live*: degrowth. However, my understanding of ecosocialism, as defined by its original proponents, was rather shallow.

After reading *For the Earth to Live*, and learning so much more of the political thinkers who have contributed to the ecosocialist project, I can confidently confirm that this is my political home. What's more, I no longer see it as an edgy political orientation, that only a few outspoken radicals could see themselves adopting. I see it quite simply as the political home of the majority of humans on planet earth. This book makes it much easier to argue for ecosocialism from a majoritarian perspective: that of the 99 per cent of humanity, and that of the rest of life on Earth.

I hope that after reading this book, you will want to declare yourself an ecosocialist as well, which, as described, is much less of a political label, and much more of an active engagement in remaking our societies in our own image.

Memories of activism

I would like to say a few words about my personal experience with Allan Todd. I was contacted by Allan in 2018, to see if I could come give a 'Green Mondays' talk at the Preston New Road fracking site, in Lancashire, just an hour north of Manchester, where I lived. And so, on a hot July day, my 6-year-old son and myself took the train up to Preston, and joined the anti-fracking Preston New Road activists, who had been camping, day and night, to disrupt and finally vanquish the first horizontal fracking site in the United Kingdom. That day was an extraordinary experience for me, and for my son. We got to see the 'coal face' of civil disobedience against the fossil fuel extraction machine: activists of all walks of life, from the Nanas, the famous group of activist grandmothers, to families with children, who came from near and far to stop the destruction of the climate, and the air, water and soil pollution from fracking operations.

My son, very much in his 6-year-old superhero phase, had no difficulties at all in understanding what the anti-fracking activists were doing. 'These are Earth Defenders, and Water Protectors, and Air Guardians,' I explained. It immediately made complete sense to him. That day was his first experience of civil disobedience, as he happily did chalk drawings on the fracking entrance road, beyond the no-entry injunction point that had recently been ordained by the High Court.

The Frack Free Lancashire campaign seemed desperate at first. Who could dream of stopping the vast power of the industrial lobbies that had convinced the UK government? Who in London would care about a place even further north than Manchester? Could an unlikely alliance of the local community, the fearsome Nanas and some travelling activists and rabble rousers, really succeed in turning the tide? In 2018, this seemed highly unlikely, and visiting the Preston New Road site felt like joining Don Quixote: here we were, armed with chalk, out on the side of the road, giving talks about the climate crisis; there they were, protected by huge fences, government support and the police, drilling away. But in 2019, after years of sustained protest (including the unjust jailing of activists by a judge who turned out to have family in the fracking supply business!), the activists won, and a moratorium was placed on fracking throughout the UK. As in most activist history, however, the activists had to remobilize in 2022, when the government threatened to reverse its decision. The battle of the water, air and earth defenders is never over.

This struggle provides a snapshot of the real-world work of Allan Todd, who has been active, a mover and shaker and bringer-together, of climate and ecological campaigns. Six years later, almost to the day, we met again at a protest against the opening of a new coal mine in Cumbria. Again, Allan had brought a large number of diverse activists together, to a remote part of the UK northwest coast. On that day, hearing about all the campaigns and efforts to stop fossil fuel expansion, we had hope and uncertainty. But again, just a few months later, the UK High Court ruled against the mine, because it had been permitted without taking into account the impacts from burning the coal.

From activism to academic freedom

I recount these experiences because they demonstrate two things. The first is that the struggle to give our living Earth a chance to keep sustaining us is the ongoing struggle of our time, and its battles are waged all over, continuously. Allan Todd is a practitioner of these battles, not just a theorist, and his writing about activism and organizing comes from a place of experience – over the years, he has sat down in front of trucks, and been arrested/convicted for actions with Greenpeace and Extinction Rebellion. He knows what it means to do the work of bringing people together, getting them to speak to each other, become active, and stay united in the face of overwhelming power and repression.

My second reason is more personal. It might seem strange for an academic such as I am to write an enthusiastic forward for a book advocating a political position. This means departing from academic 'neutrality,' and taking a firm stand on one side rather than another. I am doing this, despite knowing that the opponents of ecological action and social equity will use it against me, and even worse, against my scientist colleagues.

However, as many have pointed out, neutrality is overrated. At best, academic neutrality is a silent endorsement of the status-quo, with all its violence and inequalities; at worst, it is a cudgel used by right-wing politicians to destroy the academic freedom to research and communicate on topics of public concern, as is currently the case in the USA.

In my own trajectory, I have made full use of academic freedom, to move from physics to the study of the resource use of societies, and from there to the connections with human well-being, and from there to the political economy of unequal and destructive capitalism. I am not an ecosocialist out of wishful opinion: if anything, I am an ecosocialist because my values and my research have brought me to conclude that we must challenge neoliberal capitalism for our societies to have a chance to thrive.

If you have read this far, I hope you enjoy the book, and that it inspires you to turn your life to one of ecosocialist thinking and activism! You are in good company.

A NOTE ON USAGES

Anthropocene

This is a term – which translates as the 'Human Epoch' – suggested by many leading geologists and climate scientists as the most accurate way to describe the current situation which, they argue, is now clearly marked by signs of significant human impacts on Earth. They thus believe we are now in a different geological epoch from the Holocene Epoch, which began around 11,000 years ago. The term is usually credited to Paul Crutzen, although scientists in the former Soviet Union appear to have been using it from the 1960s onwards. An attempt in 2024 to get official acceptance of the term 'Anthropocene' was rejected by the main international geological bodies.

Austerity

Essentially, 'austerity' refers to a set of economic policies, which governments choose to apply in an attempt to deal with economic and political crises and to reduce budget deficits. Austerity usually takes the form of cutting public welfare expenditure, increasing taxes, holding down wages, and privatizing public assets. However, under capitalism – whether in liberal democratic, or in authoritarian or fascist, states – whilst public expenditure is reduced as regards social welfare spending, government funds are often re-allocated to the wealthiest one per cent, whose taxes are usually reduced.

Creeping fascism

This is a term often used in connection with the rise of modern hard-right authoritarian, far-right populist and outright fascist movements and their ideas. The term is particularly associated with the late Neil Faulkner, who saw it as covering those individuals, parties and movements which today – whilst not looking like 'first wave' fascist groups – are nonetheless taking politics in a broadly fascist direction.

Degrowth

The idea of 'degrowth' first emerged in the 1970s, and essentially argued that Gross Domestic Product (GDP – see below) should not be used as the sole measure of human and economic development. More recently, radical economists – including ecosocialists – began arguing that capitalism's drive for constant GDP-growth is causing so many increasingly-serious ecological harms, that it needs to be abandoned. It is a *positive* type of 'degrowth'which is advocated by ecosocialists (and others), based on ditching capitalism's goal of ever-increasing production of useless/harmful 'stuff'; *and*, instead, *increasing* social aspects such as health, education, housing, public services, social welfare and leisure-time. Essentially, such radical 'degrowth' is about creating an ecologically-sustainable post-capitalist/post-growth way of thriving for all.

Dematerialization

This is the overly-optimistic argument, favoured by capitalists – and by environmentalists who don't see the need to move beyond capitalism – that modernization and new technologies will eventually allow the capitalist economy to be decoupled from current levels of energy and resources use, which will thus allow the climate and ecological crises to be overcome. They believe that capitalists will develop an 'ecological rationality' that will see the need to maintain Earth's resources and ecosystem functions. As a result, those who support such views reject the need for any radical or fundamental 'System Change' to the prevailing social and economic model.

Ecofeminism

Ecofeminism is a branch of feminism, political ecology and green politics, which examines the concepts of gender and patriarchy to explore the links between humans and the natural world. The term was first used by Françoise d'Eaubonne, and now has several different strands. Today, socialist/ecosocialist ecofeminists stress the need for an egalitarian and collaborative society as the only way to restore harmony between humans and the rest of the natural world.

Ecosocialism

In essence, ecosocialism is based on the understanding that neither old-style socialism, nor a non-socialist ecology, is sufficient for establishing an ecologically-sustainable and socially-just world. Although the term was little known before the start of this century, it in fact has long historical roots which stretch back to Marx himself. Ecosocialism sees the continuation of capitalism as totally incompatible with any serious attempt to solve the climate and ecological crises.

Gross Domestic Product (GDP)
This is a monetary term, based on measuring the 'market value' of all the goods and services produced in a country. GDP is seen as indicating the strength and 'health' of a country's economy – so governments see it as essential to continually promote economic growth in order to increase GDP: something that exactly matches capitalism's need for ever-increasing production and consumption!

Labour theory of value
In simple terms, Marxists argue that, under capitalism, production/exchange of commodities is based on strict profitability, which determines the availability of products/resources – with the 'value' of commodities being based on the amount of 'socially-necessary' labour needed to produce them.

Libertaire
This term is often used to refer to the broad anti-authoritarian/ anti-bureaucratic wings of revolutionary socialist and ecosocialist movements. It is broader than 'anarchist', and covers anti-athoritarian Marxists, Trotskyists, communists and socialists – who, in many ways, are close to certain anarchist poitions when it comes to aspects of democracy and freedom. It's used in preference to the term 'libertarian' to describe such left groups, as that term is now often associasted with right-wing politics.

Mass extinctions
These are extinction 'events' which are widespread – and even global – and that see a rapid loss of biodiversity and species, significantly in excess of what's known as the normal 'background rate' of extinction. According to most scientists, there have been five mass extinctions – with the last one being 65 million years ago, which saw, amongst other losses, the disappearance of the dinosaurs. Today, many biologists believe we are now living through the Sixth Mass Extinction – this time, caused by human activity rather than natural causes.

Metabolic rift

This idea is associated with Karl Marx who, in the 1860s, argued that the economic 'logic' of the capitalist mode of production was causing a dangerous ecological breach – or 'metabolic rift' – between humans and the rest of the natural world of which humans were a part and on which their survival ultimately depended. As Marxist and socialist movements developed in the twentieth century, Marx and Engels's ecological writings were largely lost or ignored – but have been revived, from the end of the last century onwards, by scholars such as John Bellamy Foster.

Natural economy

This refers to economies in which goods are exchanged – using either direct bartering or some form of 'money' – in order to satisfy social or personal needs. The point of those exchanges is to satisfy social and human needs, rather than the accumulation of more money/wealth, and most of the goods produced and exchanged are consumed relatively locally. In some 'natural economies', if someone accumulates significantly more goods that others, such goods are shared out according to traditional customs.

Neoliberalism

This is a term to describe a political and economic philosophy which essentially seeks to return to the classical 'liberalism' associated with the early days of capitalism in the nineteenth century. Then, the prevailing view was that the state should not interfere in the operations of a 'free' market – although classical liberalism did think the state should interfere to stop working people from forming trade unions, going on strike or even having the right to vote. These ideas were given a new twist just after the end of the Second World War – but didn't really take off until the mid-to late-1970s.

Orthodox communism

'Orthodox' communism refers to the ideology that developed under Stalin, and which was subsequently promoted by Stalin's successors. This was opposed by 'classical' communism/Marxism, which drew its ideas from Marx's earlier writings (often more philosophical) and those of Marxist thinkers like Lenin and Trotsky, as well as Marx's later works.

Overshoot

In simple terms, this merely means to pass or go beyond – or 'shoot past' – a certain target or level. However, more recently, ecological thinkers such as Andreas Malm and Wim Carton have used it to describe a *deliberate* attempt to go beyond previously-agreed 'targets' – such as the 1.5°C target 'set', in 2015, as the limit to global warming by the COP meeting in Paris. In part, this is seen as 'acceptable' as it's based on having 'faith' that, at some unspecified date in the future, new technologies will be invented that will allow global heating to be reduced to safer levels. It's also an 'acceptable' option as it allows capitalism in general (and the fossil fuel industry in particular) to carry on with their 'business-as-usual' activities.

Planetary boundaries

These refer to the planetary or ecological limits beyond which it is not safe for 'human' activities to go. Originally, in 2009, Johan Rockström and his team established nine planetary boundaries, which they warned should not be breached if we wanted to maintain the stable climatic and natural conditions which have, so far, marked the Holocene Epoch. Since then, there has been mounting evidence that an increasing number of these boundaries are being breached. The danger with crossing these boundaries or thresholds is that, at a certain point, dangerous or even catastrophic changes may be triggered which humans will not be able to control.

Positive feedbacks

Essentially, these are linked to crossing planetary boundaries – and the word 'positive' is not being used in a good sense. When boundaries or thresholds are crossed – even in a relatively small way – there is the danger of triggering uncontrollable feedback loops which amplify the original disturbance, causing system instability and even breakdown.

Slow violence

This is the opposite of the 'quick violence' associated with wars, invasions, the harsh repression of resistance – and with some revolutions – and often leads to people opposing revolution in principle. However, such people often tend to ignore the 'slow violence' which occurs in societies – often over centuries – via structural or systemic 'violence' (such as economic, social and health inequalities) which results in frequent ill-health and early death for millions of people. The concept of 'slow violence' was first elaborated by the Norwegian sociologist Johan Galtung; however, the term itself was first coined by the environmentalist Rob Nixon in 2011.

Transitional reforms

These are reforms which are intended to do more than just improve the current situation in any society. Instead, the long-term intention of such reforms is to weaken the existing system *and* help prepare the way for the building of a new system. The idea of advocating such 'transitional reforms' is particularly associated with Leon Trotsky – and with the Fourth International which he established to oppose both rising fascism and the Stalinist bureaucratization of the communist movement.

Use values/exchange values

According to Marxist economics, a 'commodity' is something that's produced for sale on the 'market' for a profit. Such a commodity has both use and exchange values – but, under capitalism, it is mainly 'exchange value' which matters, as it's out of exchange values that profits are made, and thus extra capital accumulated. In the end, use values – that relate to human needs, as opposed to profits – have little importance under capitalism.

Zoonotic viruses

These are viruses or pathogens which can 'jump' or cross over from non-humans to humans. Almost 1500 pathogens can infect humans – and over 60 per cent of these are zoonotic. Whilst most animal influenza viruses rarely cross to humans, both avian and swine 'flu viruses have high zoonotic potential. Most of these viruses are associated with the growing destruction of the natural world – especially because of industrialized animal agriculture; COVID-19 is just the latest such zoonotic virus to have crossed over to humans.

INTRODUCTION: OPTIMISM OF THE WILL

In our every deliberation, we must consider the impact of our decisions on the next seven generations[1]
**The Great Law of the Haudenosaunee
of the Iroquois Confederacy**

The quotation above expresses the long-standing traditional Iroquois philosophy that formed the basis of 'The Great Law of the Haudenosaunee of the Iroquois Confederacy'. It embodies a fundamental view of the necessary relationship that should exist between humans and the rest of the natural world. Such a worldview was not unique to the Iroquois – and it's one still shared by many other Native American nations, and by other Indigenous groups around the world.

What may come as a surprise to many in the climate movement – and even to those in the labour movement – is that this was also the approach advocated by Karl Marx, who argued that, after capitalism had been replaced by a socialist society: 'the private property of particular individuals in the earth will appear just as absurd as the private property of one man in other men [slaves]. Even an entire society, a nation, or all simultaneously existing societies taken together, are not the owners of the earth. They are simply its possessors, its beneficiaries, and have to bequeath it in an improved state to succeeding generations.'[2]

This book is addressed mainly to those, in both the climate and labour movements, who feel that 'System Change' – to a society that puts people and planet first – is badly needed, but who as yet remain unconvinced that *ecosocialism* is the form such 'System Change' should take. Some readers may wish to see a 'Social Economy' replace an essentially 'Commercial Economy'; others may still think it's possible to 'green' capitalism; while others may see building a socialist (as opposed to *ecosocialist*) society, to replace one based on private ownership and profit, as the best way forward. Essentially, this book will argue that it is the fundamental philosophical approach to Nature, shared by the Iroquois nation *and* by Karl Marx, that needs to be applied, urgently and globally, if the Earth and all its inhabitants are to live *and* thrive – and that our best chance of doing so lies with ecosocialism. The book will also attempt to counter the deep – and understandable – ecological and political despair that many are currently feeling. Because, on a practical level, such despair can militate against individuals and movements taking the realistically-achievable actions needed to resist and end capitalism's destruction of Earth's climate and the natural world.

Right now, of course, it is only too-painfully clear that this ecologically-sustainable 'seven generations' approach to how humans should relate to the Earth and its resources is very far from being applied. As many people are increasingly aware – and as some of the following chapters will show – the current dominant economic relationship to the natural world is one based on short-term exploitation, destruction, spoilation and pollution. It is that approach which is driving the ever-worsening climate and ecological crises – and also the ever-widening social and economic inequalities across the world. But even though the present situation is clearly dire, and today's problems – if they remain unsolved – risk causing catastrophic collapse in the near future, it is nonetheless hugely important, in the words of that great R&B/Soul singer and songwriter Ruby Turner, to 'see horizons even in the darkest night.' It is such an outlook that helps sustain the determination, and mental and emotional strength, we need to continue struggling for a different and better future.[3]

The importance of being Gramsci

The climate and ecological crises are now so severe that we have reached a crucial stage in the relationship between humans and the rest of the natural world. It may thus seem somewhat perverse to head this Introduction with the phrase 'optimism of the will' – the second part of a political approach the Italian Marxist Antonio Gramsci urged on his comrades during the 1920s and '30s, which is often summarised as: 'Pessimism of the intellect, but optimism of the will'.

Gramsci had become General Secretary of the Communist Party of Italy (PCI) in August 1924. By then, Mussolini had been prime minister of Italy for two years, and Fascist Party gangs had been terrorising and killing members of leftwing goups and trade unions for several years. However, despite Mussolini's increasingly repressive rule – and despite Gramsci's pessimistic analysis of the current trends – he rejected a fatalistic cynicism, and instead remained optimistic about the possibility of a radical transformation to something better. In fact, as early as March 1924, in his article 'Against Pessimism', published in the PCI journal *L'Ordine Nuovo* (New Order), he had warned that 'the thick, dark cloud of pessimism… oppressing the most able and responsible militants… may in fact be the greatest danger we face at present.' Soon after, Gramsci's aphorism – 'Pessimism of the intellect, optimism of the will' – became the motto of that journal.[4]

There was certainly plenty to be pessimistic about when he wrote that article. By then, there were multiple crises of capitalism, in particular: post-war austerity and the resultant mass poverty, along with the rise of fascism. These crises were seen by many as the 'monsters' of a capitalist world-order that was disintegrating, but from which a new and better world was struggling – with considerable difficulty – to emerge. Gramsci described it thus: 'The crisis consists precisely in the fact that the old is dying and the new cannot be born; in this interregnum a great variety of morbid symptoms appear.'[5]

For Gramsci – and for many others – things would become even worse: in November 1926, he was arrested and imprisoned, remaining a prisoner until his death in 1937. During his imprisonment, the threats posed by capitalism's 'monsters' and 'morbid symptoms' increased: the 'Great Depression', the coming to power of Hitler's Nazi Party in Germany, and increasing signs of the approach of a new world war. Nonetheless – and despite his worsening health – he maintained that 'optimism of the will'. In a 'Letter from Prison', dated December 1929, he wrote: 'I'm a pessimist because of intelligence, but an optimist because of will… Whatever the situation, I imagine the worst that could happen in order to summon up all my reserves and will power to overcome every obstacle.' Today, there are still 'monsters' – those Gramsci wrote about (including the spread of 'creeping fascism' around the globe, and several nasty imperialistic wars), and now the new existential crises of climate change, ecological destruction, and pandemics. To many people today, it seems as though this twenty-first century's 'old world' is dying too – and not before time![6]

Dreaming is revolutionary!
That we need to free humanity from capitalism's 'morbid symp-toms' – in order to create another, and better, world – is be-coming increasingly clear to many. The urgent need for such a better world was underlined by the historian Eric Hobsbawm: 'If humanity is to have a recognizable future, it cannot be by prolonging the past or the present. If we try to build the third millennium on that basis, we shall fail. And the price of failure, that is to say, the alternative to a changed society, is darkness.' More recently, Neil Faulkner put it thus: 'We believe that the old order is doomed and we must build a new one based on democ-racy, internationalism, ecosocialism, solidarity with the poor and the oppressed, and the total transformation of society to serve human need not private greed.'[7]

It is the contention of this book that by far the best hope for replacing today's 'old order' with a new one along the lines Neil Faulkner envisaged lays with ecosocialism. Raymond Williams was another who, on several occasions, spoke of the necessity for hope in the struggle for a better world – for instance, the last chapter of his book *Towards 2000* was titled 'Resources for a Journey of Hope'. Yet some – including some on the left – con-sider such 'dreaming' of a better world as impractical and even unrevolutionary. The reality, however, is that such dreaming is very much 'in the spirit of Marx' – and is based firmly on the possibilities seen in present-day realities. As early as the 1960s, Che Guevara argued – against his 'orthodox' communist critics – that his dreams of the possibilities for global emancipation were *not* divorced from the existing material circumstances and con-ditions of the time. For Che, 'revolutionary hope [is] necessary for revolutionary politics and practice.'[8]

In fact, such hoping and dreaming is similar to what the Marxist philosopher Ernst Bloch termed 'Real Possible' hope (as opposed to mere wishful thinking), which 'begins with the seed in which what is coming is inherent.' Lenin, too, stressed the importance of dreaming a concrete vision of a better future. Ac-cording to him, revolutionaries 'should dream!' – even if those dreams 'may run ahead of the natural march of events.' For him, 'If there is some connection between dreams and life then all is well.' Thus, as Ernest Mandel argued: 'hopes and dreams are... categories of revolutionary *Realpolitik*.'[9]

More recently, Kate Soper has called on the left to develop 'a new political imaginary' as part of her criticisms of capitalist consumerism, and her advocacy of an 'alternative hedonism' that could make appeals for the transition to a better world seem much more attractive to people, as opposed to some rather negative ecological 'de-growth' arguments that often – and mistakenly – portray such a revolutionary transformation as essentially only being about 'giving things up' and sacrifice. Instead, she argues we need to imagine and dream – and explain – how a post-capitalist world would result in *more* pleasure, *more* fulfilment and thus *more* happiness than currently 'enjoyed' by so many, even those in work, in affluent capitalist countries. This is surely a much more positive and appealing way – 'by foregrounding the pleasures this might bring us' – to make 'the environmental and ethical case for embracing a post-consumerist (and ultimately post-growth) way of life.'[10]

For centuries, many revolutionaries, like Lenin and Che Guevara, have dreamed of abolishing exploitation, oppression, inequality, unfreedom and alienation – via the creation of a classless society. Then, however, it's been argued by some that material and cultural conditions had been insufficiently developed to allow such dreams to become reality. Thus, in that sense, such revolutionaries can indeed be seen as 'utopians', because their hopes were not fully based on the realities of their worlds – even though their rebellions were entirely justified as being part of the global struggle by humans against inhuman conditions. Nonetheless, those early utopians paved the way forward – by developing dreams and ideals, and ways of thinking and acting, that now finally make it possible to realize those hopes and dreams.

Thus, despite today's multiple crises – and the depth of many of those crises – there is more than a glimmer of hope. Because, as well as globalizing its economic reach and imposing its addictive fixation on the perpetual growth of GDP and profits, capitalism has also – inadvertently but inevitably – created a vast global army of the dispossessed and exploited, and an international environmental movement, which between them, using Marx's term, have the potential to be the 'grave-diggers' of this hugely exploitative and ecologically-destructive system. As Marx observed: 'humanity tends to set itself only such tasks as it is able to solve, since closer examination will always show that the problem itself arises only when the material conditions for its solution are already present or at least in the course of formation'. And the solution that has by far the best potential for solving those multiple crises is … ecosocialism.[11]

So: what *is* ecosocialism?

Probably the clearest and most concise explanation of what ecosocialism is, has been provided by Michael Löwy: 'The central premise of ecosocialism, already suggested by the term itself, is that non-ecological socialism is a dead end, and a non-socialist ecology cannot confront the present ecological crisis.' Essentially, ecosocialists recognize that, because of the profound crises currently facing humanity and the rest of the planet's species, *both* the socialist *and* the 'green' projects need to be redefined. These multiple and interlinked crises – climate, ecological, economic, social and political – mean that, in the twenty-first century, it is no longer simply a question of either trying to 'green' parts of capitalism, or even replacing capitalism with twentieth-century conceptions of socialism.[12]

Unsurprisingly, a political philosophy as broad and comprehensive as ecosocialism has many different strands, which emphasize different aspects of the struggle to transform today's world. Thus, within ecosocialism, there are on-going debates about some issues: such as the meaning of 'de-growth'; the relative importance of world population growth; and how to address land use and how we feed ourselves. However, the one thing ecosocialists are fully united on is that the capitalist system has to be replaced by a new system: one that puts people and planet above production and profits. Because, quite simply, there can be no truly-viable life – let alone socialism – on a dangerously-degraded planet. Given the mixed results of the COP process, it is now absolutely clear that capitalism – on any meaningful timeframe – cannot deliver that ecologically-sustainable planet.

Early eco-warriors

While many assume that the main themes of ecosocialism are a relatively-recent development, concerns about the negative impacts of capitalism on the natural world – and on the relationship of humans to it – were in fact raised by several people in the earliest days of capitalism's Industrial Revolution, which has since come to be seen as the start of what's increasingly called the Anthropocene. Some of the first to do so were William Wordsworth and William Blake.

More relevant for the development of ecosocialist perspectives, there has long been a close – though for a long time, forgotten – connection between socialism and ecology. Something still not fully understood today, even by some sections of the left. This connection goes back to Karl Marx who, in the second half of the nineteenth century, developed several key ecological ideas on the relationship between human activity – more precisely, *capitalist* activity – and nature. A succession of modern ecosocialist writers – such as John Bellamy Foster, Paul Burkett, Joel Kovel, Michael Löwy and Fred Magdoff – have convincingly established Marx's contributions.

As early as the 1850s, Marx drew attention to what he called the dangerous 'metabolic rift', or unsustainable ecological dislocation, that capitalism – because of its built-in drive for continuously-increasing production and ever-rising profits – inevitably creates between humans and the rest of the natural world. Essentially, the capitalist mode of production and accumulation treats nature as capital to be exploited, and ignores the Earth's planetary boundaries to growth. In the early 1860s, Marx also became interested in the concept of the atmospheric 'greenhouse effect', which was just being raised by the Irish physicist, John Tyndall. As has been seen earlier, Marx made the point that neither human societies in general, nor private companies in particular, *own* the natural world, and that therefore they should not degrade it. Instead, Marx argued that each generation had a duty to pass it down to succeeding generations in an improved condition.

Friedrich Engels, too, also produced ecological writings, commenting on capitalism's increasingly-destructive impact on nature, of which we are a part, and on which our survival as a species ultimately depends. As early as the 1840s, he wrote of the environmental and industrial pollution associated with capitalist manufacture and urbanization. Sadly, some of Engels's work on ecological matters – as with some of Marx's writings on such subjects – were not widely disseminated at the time. Indeed, some were not published until decades after they were written. Thus, they were largely unknown to those who nonetheless saw themselves as socialists or Marxists. Engels, for instance, warned about the possible consequences of our interference with nature, and commented that humans should not be fooled by apparent 'victories' over nature, because of unforeseen effects: 'For each such victory nature takes its revenge on us.' It is certainly possible to see the emergence of COVID-19 as an example of nature taking its revenge on us.[13]

Other theoretical contributions were made later in the nineteenth century by Edwin R Lankester (a friend of Marx) and William Morris. William Morris was particularly important, and had become concerned about environmental and ecological matters before becoming a Marxist in 1883. One of Morris's key contributions to ecological thought was to distinguish between 'wealth' and what John Ruskin – who, unlike Morris, was not a socialist – called '*illth*.' For him, 'wealth' was what nature could provide humans with, assuming they made 'reasonable use' of its possibilities. '*Illth*', however, was the harm and waste resulting from capitalism's over-exploitation and degradation of people and nature, in its relentless pursuit of profit via endless production and consumption.

Beyond the provision of decent housing, nourishment and clothing, Morris saw *real* wealth in terms of more leisure to enjoy the natural world and pursue interests, having better education and health services, and humans having the freedom to develop their creativity. Morris was also one of the first to see that capitalism made profits by promoting false 'needs', via advertizing, that led people to buy things they didn't need. He thus saw capitalism's production of endless 'stuff' – or 'the mass of things which no sane man could desire' – and the development of consumerism, as merely the production of 'illth', which diminished and degraded both humans and the natural world itself. It has been said that Morris's 1890 utopian novel, *News from Nowhere*, was 'ecosocialist in all but name… [and contained] powerful insights into the implications of the capitalist mode of production and what it had in store for the future viability of the planet'.[14]

Rachel Carson and radical environmentalism

Unfortunately, there were only a few socialist individuals and groups concerned with ecological questions in the early twentieth century. At the same time, most 'green' organizations showed little – if any – awareness of the connections between environmental degradation and the capitalist mode of production. However, soon after the end of the Second World War, some important thinkers began to emerge whose contributions in the 1950s and '60s helped lay the foundations for future ecosocialist developments. These include the revolutionary socialist Scott Nearing whose 1952 book, *Economics for the Power Age*, drew attention 'the planet's finite and declining resources, and warned of the over-exploitation and degredation of taking place.'[15]

There was also Murray Bookchin, a revolutionary leftwing green anarchist, who early on began warning that fossil fuels were having negative impacts on the climate, and 'who laid the foundations of a libertarian socialist ecology in the 1960s.' His 1964 essay, 'Ecology and Revolutionary Thought', argued that ecology should be a crucial part of any radical politics, pointing out that the competitive nature of capitalist society 'set not only each human being against the others but also the whole of humanity against the natural world.' His warnings about the likely consequences of increasing CO_2 emissions – given today's profileration of 'extreme weather events', melting ice-caps and rising sea-levels – were prophetically-accurate as regards global boiling: 'this growing blanket of carbon dioxide, by intercepting heat radiated from the earth into outer space, will lead to rising atmospheric temperatures, to a more violent circulation of air, to more destructive storm patterns, and eventually to a melting of the polar ice caps (possibly in two or three centuries), rising sea levels, and the inundation of vast land areas.'

Towards the end of that essay, however, Bookchin also drew attention – in language similar to Gramsci's – to the possibility of transformative change: 'Today, in the last half of the twentieth century, we … are living in a period of social disintegration. The old classes are breaking down, the old values are in disintegration, and the established institutions – so carefully developed by two centuries of capitalist development – are decaying before our eyes.' Ten years later, in his essay 'Towards an Ecological Society', Bookchin called for 'a fundamental, indeed revolutionary, reconstitution of society', stating quite clearly that this was because 'one might more easily persuade a green plant to desist from photosynthesis than to ask the [capitalist] economy to desist from capital accumulation.' For Bookchin, the continuous push for evermore production wasn't down to the bad intentions of any individual capitalist, but was instead down to 'the very market nexus over which he presides and to which he succumbs.' Consequently, Bookchin was convinced of the necessity to break from capitalism, and create instead an ecological society, 'not merely because such a society was desirable but because it is directly *necessary*.'[16]

However, as leading ecosocialist Alan Thornett has argued, probably 'the most important contribution to the rise of the modern environmental movement' was made by Rachel Carson. She was by far and away the most important radical environmentalist in the twentieth century, and was one of the earliest researchers and writers to warn about the growing threats to the natural world. In her groundbreaking book, *Silent Spring* – first published in 1962 – she focused on the inherent dangers in the use of organo-phosphate pesticides by large-scale agri-businesses. As she rightly pointed out: 'The balance of nature is not the same today as in Pleistocene times, but it is still there: a complex, precise, and highly integrated system of relationships between living things which cannot safely be ignored any more than the law of gravity can be defied with impunity by a [person] perched on the edge of a cliff.'[17]

Her book was a huge best-seller and became 'the single most influential ecological book of the twentieth century', bringing about mass public awareness of the fragility of eco-systems. She continued to write and campaign on ecological issues and, in *Lost Woods* – a collection of Carson's writings published after her death – there was an article from as early as 1938 in which she wrote: 'Wildlife, it is pointed out, is dwindling because its home is being destroyed. But the home of wildlife is also our home.' It is thus impossible to overstate Rachel Carson's role in the development of what eventually became today's ecosocialist movement – especially as her work made us increasingly aware of how dangerous it can be to radically alter, or even interfere with, the complex ways in which Earth's eco-systems function.[18]

Since the pioneering work of these – and other – individuals, it has become frighteningly clear that the 'ecological problem' is now this century's greatest problem, and that the world faces an existential planetary crisis. In particular, from the 1960s onwards, it became obvious to an increasing number that capitalism is ecologically dysfunctional and inherently destructive of biodiversity, and therefore needed to be replaced by a sustainable economic and social system which restores a healthy relationship with the natural world: of which we are a part and on which we depend.

To a Red-Green politics

Tragically, although there were socialist individuals and small groups who continued to propagate ecological issues in the first half of the twentieth century, most of this crucial political and economic understanding and legacy on the left was lost, as both socialist and communist parties (and their governments) increasingly focused on 'the mastery of nature' and mass production as the best ways to overcome the poverty and inequalities created by the spread of capitalism. Concerns about the environmental impact of productivism were either overlooked, or dismissed as the political concerns of the sentimental liberal middle classes.

However, a significant contributor to an ecological movement that would eventually lead to the emergence of ecosocialism was the biologist and socialist Barry Commoner. Along with Rachel Carson, he was one of the first to bring the idea of the emerging ecological crisis – and thus the need for a more sustainable relationship with nature – to the general public. His work spans the period from the 1960s to the 1980s; his first major work, *Science and Survival* (1966) was followed by his ground-breaking book *The Closing Circle: Nature, Man and Technology*, which was published in 1971. This was written from a clear ecosocialist perspective, and was followed by several more seminal works which became best-sellers – including *Making Peace with the Planet* in 1990, which was an update of *The Closing Circle*.

As Alan Thornett has pointed out, Commoner's 'Four Laws of Ecology', as explained in *The Closing Circle*, 'form the essential basis of an ecosocialist relationship between human beings and nature.' The first one is especially significant: 'Everything is connected to everything else: i.e., there is only one – intricately interconnected – ecosphere containing all living organisms, and what affects one thing affects all.' In addition, Commoner made clear the ecological dangers posed by capitalism's never-ending push for constant economic 'growth' and ever-greater profits. Above all, he drew attention to how those features of capitalism inevitably clashed with the maintenance of ecological sustainability on Earth: '[T]here must be some limit to the growth of the total capital, and the productivity of the system must eventually reach a 'no growth' condition.'[19]

Meanwhile, in the late 1970s and early '80s, people such as Edward (E. P.) Thompson, Rudolf Bahro and Raymond Williams began writing about what they called 'ecological socialism.' In 1976, Edward Thompson – himself seen as a twentieth-century romantic – published a revised edition of his *William Morris: Romantic to Revolutionary*. This massive study of William Morris, by highlighting the combination of Morris's concerns over the impacts of industrialism on nature and his revolutionary socialism, contributed to what later became ecosocialism. The continuing importance of Thompson's political and intellectual writing for ecosocialism – including his romanticism and his unfinished study of the early Romantics – for addressing today's 'planetary ecological crisis' was recently brought out in John Bellamy Foster's 2022 review of Thompson's 1980 essay 'Notes on Exterminism, the Last Stage of Civilisation'.[20]

Thompson was also an early supporter of the radical East German dissident Rudolf Bahro who, in November 1979, gave a speech in Freiburg, which was later reprinted as 'Ecology Crisis and Socialist Idea'. In it, Bahro made the importance of the connection between ecology and socialist transformation absolutely explicit: 'In the industrially developed countries today there is scarcely any subject that so urgently needs to be tackled as that of "the socialist alternative and ecology", at least no more important subject for socialists and no more important subject for ecologists or Greens.' To emphasize the existential importance of that subject, he pointed out that socialists and greens 'have long had a certain standpoint in common, *which has now become decisive for world history*. This common standpoint is a radical critique of capitalist industrialism.' Particularly relevant for debates about ecosocialism in the twenty-first century, Bahro stressed that those greens who saw industrialism in general as the problem, as opposed to *capitalist* industrialism, were simply 'misled.' Thompson later wrote an Introduction for a collection of Bahro's articles and speeches, *Socialism and Survival*, published in 1982.[21]

Raymond Williams was another who played an important role in the development of ecosocialism. In 1980, the Labour Party's ecological group, the Socialist Environment and Resources Association (SERA) – founded as early as 1973 – had published Stan Rosenthal's booklet, *Eco-Socialism in a Nutshell*, which aimed to achieve 'the blending of the traditional concerns of the labour movement with those of the fast-growing ecological lobby into a new concept: ecosocialism.' Then, in 1982, Williams wrote *Socialism and Ecology* for SERA: 'In recent years some of us have been talking about ecological socialism – ... in many countries and at a growing pace there is an attempt to run together two kinds of thinking which are obviously very important in the contemporary world.' He went on to point out how many people in the early Industrial Revolution – including at least one of its key inventors and engineers – had described how industrial processes were already disturbing and driving out nature. Then, after criticising modern socialism for continuing to see more production as the only way to cure poverty and thus 'an absolute human priority', he also urged the need to move away from the idea of a 'non-political' ecology. In place of both these political strands, he argued for the construction of an international and 'ecologically conscious socialism.'[22]

Ecosocialism rising!

Other crucial strands in the red-green thinking that eventually fed into the development of ecosocialism are ecological feminism, or ecofeminism; and a growing awareness in the global North of the importance of struggles by Indigenous peoples to protect their lands from capitalist destruction. Ecofeminism sees the devalueing, and the resulting degradation, of nature as arising in large part from the same dualistic Cartesian philosophy that also results in patriarchy and the oppression of women: the linked ideas that Nature is inferior to 'Reason', and that women are 'close to nature' and thus 'inferior' to men who are said to be 'rational'. Then, in the 1990s, ecological socialists became increasingly aware of the struggles of Indigenous peoples in the global South for land rights and against environmental destruction in their countries.

Finally, the emerging ecosocialist agenda was also given a boost by the formation of the UN's Intergovernmental Panel on Climate Change (IPCC) in 1988, which started a process of international summits that at least began to acknowledge that there *was* a problem with greenhouse gas (GHG) emissions in the developed world, and their increasingly negative impact on the climate. There were also numerous collective statements from the global South, including the Mount Tamalpais Declaration of 2000, and the People's Agreement of Cochabamba in 2010. These called for 'decolonization' of the atmosphere, and pointed out the injustice of 'green' mechanisms, such as carbon trading, as they were failing to deal with the debt that the rich North owes the South.

One example of how wider sections of the left began embracing ecosocialism as a way to deal with the climate and ecological crises is provided by developments within the Trotskyist Fourth International. As early as 1991, it had initiated a debate around 'Socialist Revolution and Ecology', and this had continued until its Sixteenth World Congress in 2010, when a 2009 proposal by its International Committe – to identify as an explicitly 'ecosocialist' organization – was overwhelmingly accepted.[23]

Then, at its Seventeenth World Congress in 2018, it overwhelmingly passed the resolution: 'The capitalist destruction of the environment and the ecosocialist alternative', prepared by its Ecology Commission. This recognised that: 'The struggle to defend the planet and against global warming and climate change requires the broadest possible coalition involving not just the power of the indigenous movements and the labour movement but also the social movements that have strengthened and radicalized in recent years and have played an increasing role in the climate movement in particular.'[24]

As the impacts of the climate and ecological crises have become more apparent in the new century, there has been increased discussion on the broader left about what 'ecological socialism' actually entails. The debate has moved significantly from the prevailing views of the 1980s, when many socialists thought their organizations merely needed to be more aware of environmental issues. Increasing awareness of the scale and worsening impacts of the climate crisis has led to a recognition that politics needs to be a close combination of red and green issues – and a growing acceptance that it is no longer possible to be simply a socialist, or even a Marxist, without taking into account how the natural environment is fundamental to the reproduction of humanity. As a result, books that took Marxism as their starting point, began seriously to address the ecological issues, and to develop clearer and fuller definitions of ecosocialism.

At the same time, there has been a growing international convergence on the left, in both the global North and the global South, of the climate and ecological struggle *and* the class struggle against neoliberalism and its rapidly-increasing inequalities across the world. The formation of the World Social Forum (WSF), and the production of a first *Ecosocialist Manifesto* in 2001, set out a clear and consistent ecosocialist position. In 2009, the Ecosocialist International Network was set up at the WSF meeting in Belém (Brazil) from which came the

Belém Declaration, essentially a second version of the earlier *Ecosocialist Manifesto*.

Yet there are still those on the left who tend to dismiss these ecological concerns as either a form of conservative neo-Malthusian politics, a romantic rejection of industrialization, a distraction from the class war, or simply as elite preferences and a desire to carve out some privileged professional position in society. At the same time, another problem is that 'green' political groups and social movements often still associate *any* form of socialism with a purely 'materialist' outlook, and a blind commitment to endless 'productivism' and growth. The poor environmental records and performance of past social-democratic/labour parties and governments in Western Europe, and the often-appalling environmental records of the former Soviet Union and the other so-called socialist or post-capitalist economies in Eastern Europe, also present a stumbling block to a much wider acceptance of ecosocialism within the climate movement. Initially, this has resulted in organizations such as Extinction Rebellion arguing that the movement should be 'a-political'. Such a position has resulted in many actvists being unwilling to discuss precisely what 'System Change' is needed in order to avoid 'Climate Change'.

This is why there is an increasing awareness amongst part of the left that the way forward for the ecosocialist project is to acknowledge the poor ecological record of twentieth-century socialism, and instead to go back to Marx and the other nineteenth-century pioneers. From there, many ecosocialist theoreticians have developed a credible alternative – focused on social and ecological wellbeing – to constantly-increasing and unsustainable capitalist 'growth'. Currently, there is an increasing general understanding – especially amongst young people – of the causes and impacts of global boiling, accompanied by a growing awareness of the destructive nature of capitalism as it has spread across the world, especially in its most recent global neoliberal form.

Reasons to be hopeful
Before the start of this century, there were not many radical or revolutionary left organisations in the UK that specifically identified as ecosocialist. However, since the COVID-19 pandemic,

there has been a definite shift towards more left groups embracing ecosocialism. Firstly in 2021, an Ecosocialist Alliance was formed in which several ecosocialist groups co-operated to produce joint statements in preparation for G8 and COP meetings. Most recently, 2023 saw the launch of an 'Ecosocialism Conferences' project.

While it's difficult at times to maintain hope – especially when some things are going backwards rather than progressing – it is essential to maintain Gramsci's 'optimism of the will', and to dream of the real possibility of creating an ecologically-sustainable and economically-just world. As more and more people come to realize that 'capitalism has outlived its usefulness', it's also important to hang on to the fact that we *don't* have to continue living like we are right now – capitalism is *not* a 'forever' system. While many of the following chapters may, initially, merely confirm some in their depression about the chances of altering course, the hope is that, by the end of the book, most will see that such hopes and dreams are *not* utopian or unrealistic. On the contrary, those hopes and dreams – along with a determination to keep on struggling – are essential 'For the Earth to live'. In fact, if he were alive today, Ian Drury – who has been described as a 'voice for the disenfranchised' – would (probably!) be singing that there *are* 'Reasons to Be Hopeful'.[25]

THE CLIMATE SHIT WE'RE IN

*Right now, we are facing a man-made disaster of global scale,
our greatest threat in thousands of years: climate change.
If we don't take action, the collapse of our civilisations and the
extinction of much of the natural world is on the horizon.*[1]
David Attenborough

The title for this chapter was in part suggested by Bill McKibben's 2005 article 'The debate is over', which identified three eras as regards growing awareness of the climate crisis: firstly, there was the 'I wonder what will happen?' era; followed by the 'Can this really be true?' era. According to him, by 2005, we had already entered the 'Oh shit!' era. What particularly concerned him was the uncertainty about the severity of what would happen, given that 'the world... has some trapdoors – mechanisms that don't work in straightforward fashion, but instead trigger a nasty chain reaction'[2]

Yet, in April 2024, Chris Stark – the outgoing head of the UK's Climate Change Committee (CCC) – made it clear that we still face an uphill struggle to get the politicians to take the necessary actions. According to him, the decision by Sunak's Tory government to 'dilute' key green policies had set the UK back as regards achieving net zero. He said: 'I think we have moved from a position where we were really at the forefront, pushing ahead as quickly as we could on something..., fundamentally beneficial to the people living in this country, whether you care about the climate or not...We are now in a position where we're actually trying to recover ground.'[3]

However, the climate crisis is only one of many problems facing humanity this century. There are so many serious crises that it's possible to say, given where we're at right now, that the 'world is on fire', in what has rightly been described as 'the worst crisis in human history', in which we face multiple threats resulting from a 'compound crisis of the world system.' According to Michael Albert, since 2020, 'words like "permacrisis" or "polycrisis" have become common currency, reflecting a broadening awareness that ours is an age of interconnected systemic crises with no clear end in sight.' These systemic crises include eight particularly serious ones: the climate crisis; the ecological crisis; threats from pandemics; rising economic and social inequalities; the erosion of public health and welfare provision; the worldwide erosion of democracy; the rise of far-right populism and of outright fascist movements; and an increasing number of imperialist conflicts that could possibly lead to a nuclear Third World War. As regards the latter, especially worrying are Israel's genocidal attacks on the Palestinians in Gaza, Lebanon and Iran; the Russian invasion of Ukraine; and the US and its NATO allies currently identifying an 'authoritarian axis' – comprising Russia, China, North Korea and Iran – as a threat to 'Western democracy'.[4]

The Introduction to this book concluded with these words: '… there are "Reasons to be Hopeful."' This is despite the fact that, as 2024 – and this book – began, there seemed to be very few reasons to justify such Gramscian optimism. Instead, with multiple global crises that are getting worse as every year goes by, 'pessimism of the intellect' seems to be a much more realistic attitude to adopt. For if these crises are not radically and quickly addressed, the result will almost certainly be the collapse of human civilization as we know it – as David Attenborough starkly noted back in 2018 at COP24.

Of all those threats, three in particular stand out: a climate crisis that threatens to seriously undermine food production, and make many areas of the Earth uninhabitable; an ecological catastrophe that will result in mass death for many Earthlings – including humans – from the destruction of ecosystems and increased risks of deadly pandemics; and the 'hollowing out' of democratic political systems around the world (ushering in a period which has been described as the 'twilight of democracy'), along with the increasing spread of 'creeping fascism'. This chapter will deal with just the climate crisis – the ecological crisis, the threat of new pandemics emerging, and the various aspects of the political crisis – will be dealt with in subsequent chapters.

As will become increasingly obvious, many – if not all – of these global twenty-first century crises – and the generally inadequate responses to them, are closely related. Recently, this has been referred to as an 'imperial mode of living' based on globally-unsustainable systems and 'practices of daily living, such as manufacturing and consumption … as well as hegemonic, internalized ideas of a "good life" and of societal development.'[5]

The age of stupid

Maintaining any justifiable 'optimism of the will' first of all requires a 'pessimism of the intellect' that fully takes on board just how bad things really are as regards the ever-worsening climate crisis – and how the world's governments are still failing to take the right steps, despite the overwhelming evidence from an increasing number of scientific reports.

So, as a start, it may be worth recalling a line from the 2009 film 'The Age of Stupid': 'Why didn't we save ourselves when we had the chance?' Just over a decade later, in 2021, Professor David King, the former Chief Scientific Advisor to Blair's and then Brown's Labour governments, said: 'What we do over the next three to four years, I believe, is going to determine the future of humanity. We are in a very very desperate situation.'[6]

King had previously advised the UK government that limiting global warming to 1.5°C was vital to avoiding uncontrollable environmental changes in the polar icecaps and Himalayas, but said he now realized this was wrong: the crucial point had already been crossed. Another stark warning comes from David Wallace-Wells, who has painted a picture of life on Earth – involving social and economic collapse – at 2°C of global boiling which, in 2019, was already coming to be seen as an optimistic goal: 'The ice sheets will begin their collapse, 400 million more people will suffer from water scarcity, major cities in the equatorial band of the planet will become unliveable, and even in the northern latitudes heat waves will kill thousands each summer.' Yet, as Mark Lynas argued in 2020, if the current 'business-as-usual' emissions path was maintained, we could see: 'two degrees as soon as the early 2030s, three degrees around mid-century, and four degrees by 2075 or so.'[7]

Words that are increasingly used in relation to the rising number of extreme weather events (which are definitely *not* 'natural' disasters) are 'unprecedented' and 'record-breaking'. The fact that they are used so frequently – rapidly becoming the new norm – is a warning that has been ignored by many. Because so many events linked to the climate crisis happen each year now, it's easy to focus just on the current event, forgetting those of the very recent past. The years since 2020 have been particularly instructive as to what we can expect: not just in the global South, but increasingly in the global North as well – including the UK. In 2021, global sea levels set a new record high – and climate experts predicted that these would continue to rise. In the summer of 2022, the Environment Agency warned that 200,000 UK homes were going to be lost to rising sea-levels by 2050; while the Meteorological Office issuing two heat health-warnings – during that summer, the UK recorded its highest ever temperature: 40.3°C, resulting in 2,985 excess deaths from the heat. Additionally, wildfires destroyed over 60 homes in parts of London, the South-East, East Anglia and Yorkshire; while a serious drought adversely affected crops in several areas. For many other parts of Europe, summer 2022 – called a 'heat apocalypse' – was even worse, with severe droughts and four times as many wildfires as the historical average.

Worldwide, the summer of 2022 broke records for the number of climate disasters: in East Africa, for instance, millions faced starvation because of severe and prolonged droughts. By the time COP27 had begun in Sharm El-Sheikh in November 2022, there had been devastating floods in Pakistan which had directly impacted 33 million people, as well as floods in Australia and elsewhere. Other climate disasters that year had seen wildfires, extreme heat events, ice melt, drought and extreme storms on many continents.

In October 2023, Storm Babet – an intense extra-tropical cyclone – hit large parts of northern Europe. This included the UK, with the Environment Agency reporting 7 dead, over 1,250 homes flooded, hundreds more families forced to evacuate their homes, over 30,000 properties needing extra flood protection, and over 40,000 homes left without power. Additionally, DEFRA reported on the likely impacts of those floods, such as the rotting of potatoes and other crops – underlying just how vulnerable food supplies are to this ever-increasing number of extreme weather events.

The 'State of the Climate' Report
On 24 October 2023, whilst clear-up operations from Storm Babet were still taking place, the 'State of the Climate' Report – drawn up by over 15,000 eminent climate scientists – was published. This highlighted the suffering caused, across the globe, by record-breaking climate extremes – and raised alarms about the possibility of widespread societal and ecological collapse in the future: something David Attenborough had warned about at COP24 five years before.

As well as criticising recent *increases* in subsidies to the fossil fuel industry, which is the primary driver of climate change, the Report stated that: 'The truth is that we are shocked by the ferocity of the extreme weather events in 2023. We are afraid of the uncharted territory that we have now entered.' The Report pointed out that 2023 had already been a particularly devastating year of extreme wildfires, floods, and heatwaves – all linked to, and amplified by, climate change. The scientists listed 14 record-breaking 'Climate-Related Disasters' that had occurred since November 2022 – and also pointed out that the list did not include every such disaster. Many of those listed – like Storm Babet – increasingly occurred in the global North. These included heavy flooding in parts of the US (over 20 fatalities and $3.5bn worth of damages to property); intense wildfires in Canada (10 million hectares of forests destroyed, 30,000 people forced to leave their homes); deadly heatwaves in southern and mid-western areas of the US (almost 150 deaths); the deadly Cerberus heatwave in Europe in July 2023. However, 2023 had already seen many more climate-related disasters in the global South – these included record-breaking and deadly temperatures in Asia, and extreme storms and floods in many countries (including Japan, China and Libya), that had led to thousands of deaths, massive destruction of homes and infrastructures, and food shortages resulting from the disruption of farming and loss of crops. The September floods in Libya alone had killed over 10,000 people.[8]

The Report's authors suggested that temperatures in July 2023 may well have been the warmest on Earth over the past 100,000 years, which they saw as: 'a sign that we are pushing our planetary systems into dangerous instability.' The Report also pointed out how the climate crisis – and, in particular, 'global boiling' – has deepened since the start of this century. After pointing out that, prior to 2000, global daily mean temperatures had never been more than 1.5°C above pre-industrial levels, and had only occasionally exceeded that number since then, the Report noted that: 'However, 2023 has already seen 38 days with global average temperatures above 1.5°C by 12 September – more than any other year – and the total may continue to rise.' The Report saw the effects of such 'global boiling' as becoming progressively more severe – running the real risk of a worldwide societal breakdown, as a result of unbearable heat, frequent extreme weather events, food and freshwater shortages, rising seas, more emerging zoonotic diseases, and a resulting increase in social unrest and geopolitical conflict. The scientists actually warn that: 'By the end of this century, an estimated 3 to 6 billion individuals – approximately one-third to one-half of the global population – might find themselves confined beyond the liveable region, encountering severe heat, limited food availability, and elevated mortality rates because of the effects of climate change.' Finally, the authors – who have spent years monitoring 35 of Earth's 'vital signs' (e.g., global tree cover, greenhouse gas concentrations, ocean temperatures, and livestock populations) – pointed out that 20 of those 'vital signs' are now at record extremes: up from 16 in 2022. Their conclusion was a clear one for the entire global climate movement: 'As we will soon bear witness to failing to meet the Paris agreement's aspirational 1.5°C goal, the significance of immediately curbing fossil fuel use and preventing every further 0.1°C increase in future global heating cannot be overstated.' That particular estimate proved to be – like so many other climate predictions – to have been an underestimate: as revealed in more detail below, we now know that in fact 1.5°C was exceeded *every* day in 2023.[9]

The new 'normal'
Also in 2023, the Global Tipping Points Report had warned about the increasing risks of 'irreversible change' as regards both climate change and nature loss. This Report pointed out that of the Earth's major tipping systems, five were already near to crossing irreversible tipping points, and that – in words echoing David Attenborough's 2018 warning – this posed 'threats of a magnitude never faced by humanity.' Their Report stressed that these threats were likely to occur 'within decades' and, particularly concerning, would happen 'at lower levels of global warming than previously thought.' One particularly serious threat was a 'global-scale loss of capacity to grow major staple crops.' More generally, triggering just one Earth System tipping point 'could trigger another, causing a domino effect of accelerating and unmanageable damage.'[10]

That same year, the UN warned that not enough was being done; and in July 2023 – with the world experiencing an unprecedented number of record-breaking extreme weather events – UN Secretary-General António Guterres upgraded global warming to 'global boiling.' Then, in January 2024, the EU's Copernicus Climate Change Service confirmed that sense of urgency we so badly need, when it announced that their earlier predictions – that 2023 would be the warmest year on Earth since records began – had been confirmed. For the first time, across the entire year, *every day* had exceeded 1.5°C, compared to pre-industrial levels. Technically, though, that doesn't mean we have now definitively passed the 1.5°C threshold – for that to happen, that would have to be the average increase over a period of twenty years or more. However, as Malm and Carton warn, just one year of such an increase could be 'a springboard' which would see the world 'fly over the boundary.' In fact, as early as 2022, the World Meteorological Organization (WMO) had said that there was a 50 per cent chance that the planet would exceed 1.5°C in at least one of the next five years – and that if that happened, global warming 'between 2022 and 2026 could reach up to 1.7°C; come 2028 or 2030 and it could be higher still.' In January 2024, Copernicus reported that that month was the warmest January ever, even beating January 2023 – and that February was heading towards being the warmest February ever![11]

If that isn't enough to convince people that we truly are in some real bad climate shit, further reports in April 2024 painted an even gloomier picture. The UN's Climate Chief warned we have just 2 years left in which to prevent catastrophic problems, pointing out that current plans – assuming they are actually implemented – will barely cut emissions by 2030. That same month also saw worrying reports of what had happened during 2022 in Antarctica. In the coldest place on Earth, scientists had recorded a temperature rise of 38.5°C *above* its seasonal average: the largest jump in temperature ever measured anywhere on Earth. This temperature-leap – which wasn't an isolated one in this area – was described as 'simply mind-boggling' and raising fears of catastrophe, as 'there is now a real danger that some significant sea level rises will occur in the next few decades.' If the ice sheets and glaciers of west Antarctica continue to shrink, it is likely that sea levels could rise by as much as five metres – significantly higher than predicted previously by the IPCC, which saw rises being no higher than about one metre. Scientists now think the IPCC has 'very probably underrated by a considerable degree the threat that now faces humanity.'[12]

Further evidence of the worsening multiple crises emerged that same month. In the UK, the National Farmers' Union (NFU) issued an urgent warning about likely food shortages because of several storms which had hit the UK after Storm Babet. According to the NFU, torrential downpours – in particular, during Storm Henk in January 2024 – and 'unseasonably low spring temperatures' increasingly meant produce 'simply doesn't leave the farm gate.' As a result, farmers had lost crops over the winter and were now unable to plant spring crops: the result was that 'around 40 per cent of the [wheat] crop in the country [was] in a poor or very poor condition.' Then, in mid-April, Dubai in the United Arab Emirates was hit by 'apocalyptic floods' following four large storms over two days. This extreme weather event came just five months after what could be described as hubristic comments made by Sultan al-Jaber in November 2023, just before becoming president of COP28, that there was 'no science' behind demands to phase-out fossil fuels and that doing so wouldn't allow sustainable development 'unless you want to take the world back into caves.' It also emerged that the UAE had planned to use its role as COP28 host to make new oil and gas deals. Though after an international backlash, he later 'clarified' what he had really meant![13]

In fact, virtually every year of this century has made it increasingly obvious that we're currently living through the greatest crisis in human history: a crisis consisting of several unprecedented but linked crises that are creating dangerous – and possibly fatal – ruptures in the Earth System. If those crises are not dealt with in sustainable and healing ways, there is a real risk that, in the most extreme apocalyptic scenario, we will see the extinction of huge numbers of species on this planet, along with the deaths of millions – and possibly billions – of humans.

The carbon count

The Climate Crisis is essentially one of an increasingly-warming world – itself the result of increasing concentrations of greenhouse gases in the atmosphere. This is particularly associated with carbon dioxide (CO_2) which, before the Industrial Revolution, stood at 280 parts per million (ppm) – a level that had been fairly constant for around 10,000 years before 1750: a period during which human civilisation had taken off. That this might be a potential problem was first noted by Jean-Baptiste Fourier as early as the 1820s.

However, in 1856, Eunice Foote, an amateur scientist in the USA, was the first to suggest that an increase in CO_2 would result in a warmer planet. This was more fully developed by John Tyndall, an Irish physicist who, it seems, was unaware of her findings. In 1859, he began studying how the atmosphere regulated Earth's temperature. He concluded that certain gases – such as carbon and methane – 'absorbed heat via infrared radiation' and that, as it was transferred to the air and surface, it had what we now call a 'greenhouse effect': which, by trapping some of that heat, helped 'warm the earth…to create a habitable climate… Thus the atmosphere admits of the entrance of the solar heat; but checks its exit, and the result is a tendency to accumulate heat at the surface of the planet.' From all this, he correctly concluded that any change in the proportion of gases – such as CO_2 – in the atmosphere could change the climate.[14]

Tyndall's work was taken further by August Arrhenius, a Swedish scientist, who argued that fluctuations in the concentration of carbon in the atmosphere altered the amount of water vapour in the air, and that this helped explain changes in Earth's temperatures. In 1896, he revealed how such changes, as well as helping to trigger ice ages, could also warm the Earth beyond average temperatures. In particular, he argued that humans were already helping to increase temperatures on Earth because of the release of massive quantities of carbon into the atmosphere as a result of an economy heavily based on the burning of fossil fuels.

However, at a time when industrial capitalism was still expanding and had not yet spread to today's levels of intensity and global reach, he thought that it would take centuries before humans had doubled CO_2 atmospheric concentrations. Yet, since then, there has been a fifty per cent increase: in 2010, it was 390ppm – already higher than ever before in recorded human history; by 2022, with an average growth of 2ppm each year, this had reached 421ppm; by 2023, it was 423ppm; and, at the beginning of 2024, a further rise meant it was standing at 425ppm. Yet Hansen and others have made it clear that it is imperative to keep CO_2 concentrations to 350ppm, in order 'to preserve a planet similar to that on which civilization developed and to which life on Earth is adapted.'[15]

Both Marx and Engels studied the work of contemporary natural scientists – and it's known that Marx attended several of Tyndall's lectures and was extremely impressed by his conclusions regarding solar radiation and its impacts on Earth's temperature. And it was the work of Tyndall and Arrhenius that would ultimately lead to the development of scientific work on global warming from the 1960s onwards. Since then, we now know that, over long geological times, there have been significant changes in the atmospheric concentrations of such greenhouse gases. What is particularly concerning today is that we are seeing changes in these concentrations that are unprecedented for human history, combined with overwhelming evidence that humans and socio-economic forces – identified by ecosocialists as the global capitalist economy – are continuing to drive those concentrations upwards. As Julia Steinberger has pointed out: 'The climate crisis is brought to us by highly unequal and undemocratic economic systems…the climate crisis is a crisis of wealth accumulation.' Such a conclusion is confirmed by an article which appeared in the journal *Nature Sustainability*, in September 2022. According to Lucas Chancel, in his paper 'Global carbon inequality 1990-2019', the top 10 per cent were responsible for almost 50 per cent of global GHG emissions in 2019 – while the bottom 50 per cent of people emitted just 12 per cent of such emissions. Furthermore, he found that, since 1990, 'the bottom 50 per cent of the world population [had] been responsible for only 16 per cent of all emissions growth.' Over the same period, 'the top 1 per cent [had] been responsible for 23 per cent of the total.' The high figures for the one per cent were down to a combination of their investments and their lavish lifestyles and consumption patterns.[16]

Code red for humanity

In 2021, UN Secretary-General António Guterres warned that the latest IPCC Working Group Assessment Report, which drew attention to the rapidly-worsening and rapidly-accelerating impacts of global heating, was 'a code red for humanity.' The IPCC Working Group predicted that the much-vaunted 1.5°C 'limit' for the rise in the average global temperature, the stated aim since COP21 in Paris in 2015, would be breached by 2040 if 'business-as-usual' continued. One of the Report's most alarming aspects was that it showed that the harmful impacts of global heating were now arriving much faster – and more severely – than had previously been predicted. [17]

In the five years before Guterres issued his 'code red' warning, there had been frequent massive wildfires in Greece, Australia, the US, Canada, and even in the Arctic Circle. These frequent wildfires – often the result of 'unprecedented' devastating multi-year droughts – along with other extreme weather events, were linked to changes in atmospheric circulation, possibly triggered by the disappearance of Arctic Sea ice. In addition, many parts of the world experienced devastating floods, not just in Asia but even in Western Europe. As Christian Zeller has pointed out, such extreme weather events are now the 'new normal' and are occurring 'four to five times as often as in the 1970s and [causing] seven times as much damage.' The most recent IPCC Assessment Reports have actually pointed to capitalist growth, capitalist inequalities and the responsibility of the richest sections of humanity, as some of the prime factors in GHG emissions –something ecosocialists have been pointing out for at least the past 20 years.[18]

Well before then, plenty of climate writers and activists had been warning us for years that we needed to start worrying and, more importantly, *acting*, if crucial tipping points and unwelcome positive feedbacks were to be avoided. In 2014, Naomi Klein had drawn attention to how, ultimately, humans are dependent on the natural world; and how extreme weather events show how vulnerable we really are. While, in 2020, Mark Lynas, in *Our Final Warning: Six Degrees of Climate Emergency*, had warned how we risked going above a 1.5°C increase in the average global temperature. Even the UK's Climate Change Committee (CCC) has said the world could be on course for a 4°C increase. In fact, 'just' a 3°C rise would lead to colossal firestorms destroying the Amazon rainforest, thus pouring tens of billions of tonnes of additional carbon into the atmosphere.[19]

Yet the evidence shows that GHG emissions are continuing to rise and that, at present, no major country is taking the climate actions needed. In fact, oil-producing countries and fossil fuel companies are actually planning to *increase* production significantly between now and 2045. All this is putting Planet Earth on a path to catastrophic climate breakdown. As a result, atmospheric concentrations of greenhouse gases are higher than at any time in past 800,000 years or more. While the global surface temperature has increased faster since 1970 than in any other fifty-year period over the past 200,000 years or more. As a result of global heating, sea levels are rising – and will rise by over half a metre this century if GHG emissions continue at current levels; while the planet's ice-caps and glaciers are melting at unprecedented rates. As a result, the conditions of everyday life for billions of people are rapidly deteriorating – and forcing many to become climate refugees.

Before leaving the issue of climate behind, it's worth bearing in mind that the results of COP28 were rather like the proverbial curate's egg: only good in parts! Whilst it was good that the signatories recognised the need to get off fossil fuels (not just coal, as in COP26 in Glasgow), there was no really serious programme of action to halt the mounting climate and ecological crises. And 'crisis' is no exaggeration: in May 2023, the UN reported that, between 1970 and 2021, extreme weather events – increasingly supercharged by global boiling – had caused over 2 million deaths, as well as costing over $4.3 *trillion* in economic losses. Since then, as we've seen, 2023 proved to be the worst year – so far! – for the number of extreme weather events, and shattered climate records, to officially become the hottest year on record. As a result, the earlier '1.5°C to Stay Alive!' 'target' seems to have been already been passed. This is why – with GHG emissions continuing to rise – the UN warned, just before COP28, that we are now heading for what António Guterres has described as a 'hellish future' of at least 3°C of global boiling. According to the UN Environment Programme Report, even a rise to 2.5°C would still be a 'catastrophic scenario'.[20]

According to one Climate Action Tracker estimate in 2023, even if countries actually acted on the pledges made, average global temperatures would still rise to 2.5°-2.8°C above pre-industrial levels by 2100. However, if countries continue following current policies and levels of CO_2 emissions, by 2100, we risk seeing an increase in average global temperatures of 2.8°-3.2°C. Whilst, if countries do not act at all to reduce emissions, the average temperature could rise as high as 4.1°-4.8°C. However, although an update issued in December 2023, after the pledges of COP28, had revised the earlier predictions down, the worst-case scenario still saw global boiling possibly reaching as high as 2.9°C above pre-industrial levels.[21]

Then, in September 2024, came reports that showed August 2024 was Earth's hottest August in 175 years of climate records – and was also the '15th month in a row of record-warm months' in the Northern Hemisphere, making summer 2024 'the warmest meteorological summer on record', hitting an average temperature 1.52°C above pre-industrial levels. While, for the same period (June to August), the Southern Hemisphere experienced its warmest winter on record. Globally, the period from January to August 2024 ranked as the warmest ever recorded, at some 1.28°C above even the twentieth-century average. These figures led the National Centers for Environmental Information's 'Global Annual Temperature Outlook' to conclude that 'there is a 97% chance that 2024 will rank as the world's warmest year on record.' Not surprisingly, such temperatures resulted in a large number of 'significant climate anomalies and events' around the world that month – including severe floods, hurricanes, and melting polar ice – resulting from record global land and ocean surfaces temperatures. These severely impacted countries in both the global South and the global North. Then, in October 2024 – with record air and ocean temperatures in the Gulf of Mexico – parts of Florida and the Yucatan Peninsula were devastated by Hurricane Milton, which reached Category 5, with winds reaching up to 180mph. This came less than two weeks after Hurricane Helene had hit the same areas. Yet, as David Attenborough warned way back in 2011, in the last of his *Frozen Planet* programmes, possibly the biggest threat to humanity from global boiling is the rapidly-increasing loss of Arctic and Antarctic polar ice. Because disappearing ice-caps will no longer reflect back solar heat, the albedo effect will actually increase global boiling by making the oceans warmer. It will also lead to a general rise in sea-levels by least one metre. That would threaten the homes and livelihoods of millions on people currently living in coastal areas.[22]

However, if all the above hasn't provided enough reasons to be experiencing 'pessimism of the intellect', don't despair (pun intended!), the next chapter, which focusses on the mounting ecological damage being done to Earth's life-support systems, will provide even more evidence of the fatal and widening metabolic rift which has been created between humans and the rest of the natural world – and thus further reasons why we need to step up the struggle to bring about a radical and transformational System Change as quickly as possible.

THE SIXTH MASS EXTINCTION

*A deep chasm has opened up in the metabolic relation
between human beings and nature – a metabolism that is
the basis of life itself. The source of this unparalleled crisis
is the capitalist society in which we live.*[1]
John Bellamy Foster

As serious as the implications of the climate crisis are, arguably the
biggest threat may come from the ecological and biodiversity crisis,
which is already having catastrophic impacts on the Earth's support
systems. It is now generally accepted that we are currently living
through the Sixth Mass Extinction. The scale of the ecological threat
confronting us was starkly summarised in 2010: 'It is now universally recognized within science that humanity is confronting the prospect – if we do not soon change course – of a planetary ecological
collapse.' All three of the levels in which life is organised are now under threat: ecosystems, species, and genes. According to eminent biologist Edward Wilson, the increasing impact of humans on Earth's
systems today has 'the potential destructive power of the Chicxulub
strike.' The Chicxulub Strike was the massive asteroid which, around
65 million years ago, hit the coast off Mexico's Yucatán peninsular –
leading to the extinction of the dinosaurs, along with about seventy
per cent of all species. That mass extinction event was the fifth in
the planet's history – and like the other mass extinctions, evidence
suggests that 'Earth required roughly ten million years to recover.' As
he points out, 'In most extinctions, … the causes are multiple, linked
to one another in some way… As extinction mounts, biodiversity
reaches a tipping point, at which the ecosystem collapses.'[2]

Planetary boundaries

Of the two 2024 reports previously mentioned, the most worrying
one is the Global Tipping Points Report, as this touches on so much
more than just climate change: the dangers arising from breaching
planetary boundaries. The concept of planetary boundaries was
developed by a group of scientists, led by Johan Rockström at the

Stockholm Resilience Centre – also involved was James Hansen, the USA's leading climatologist. They identified nine planetary boundaries relating to human activities on Earth, and safe limits for each of those. Going beyond those safe limits risks the environment not being able to self-regulate anymore – thus undermining the stability of the Holocene: the period in which human society developed. They were particularly concerned about the impacts of the Industrial Revolution. According to them, 'transgressing one or more planetary boundaries may be deleterious or even catastrophic due to the risk of crossing thresholds that will trigger non-linear, abrupt environmental change within continental-scale to planetary-scale systems.'[3]

In 2009, these scientists noted that three had already been crossed: climate change, biodiversity loss and the nitrogen cycle. Their findings were updated and refined in 2015, with them concluding that four 'safe-zones' had now been breached: biosphere integrity, climate change, land-system change, and the biogeochemical cycles (nitrogen and phosphorus). The change in wording from 'loss of biodiversity' to 'change in biosphere integrity' emphasizes that not only is the number and variety of species important for maintaining the stability of the Earth System, but so is the functioning of the biosphere *as a whole*. In 2022, the 'chemical pollution' boundary (now called 'introduction of novel entities', to encompass all the main types of human-generated materials that disrupt Earth System processes) had been crossed; and in 2023, freshwater change became the sixth planetary boundary to be transgressed.

An interesting – and extremely useful – development of this idea of planetary boundaries was made in 2012, by the economist Kate Raworth. Following on from the discipline of ecological economics which developed in the 1980s, she added a crucial element: *social* boundaries – stressing that, as well as environmental or planetary boundaries being important, so too are social and economic boundaries for people. The important thing was not 'economic development' per se – but *sustainable* development that dealt with environmental *and* social needs: the latter including life's essentials like access to healthcare, education, housing and greater economic equality.

She published her findings in 2017, in *Doughnut Economics*. Her environmental and social 'doughnut' presents a very different economic model from that presented by mainstream neoliberal economists who stress the importance, desirability and possibility of constant and constantly-increasing economic 'growth'. An approach they share with those very corporations whose unchecked desire for ever-increasing profits is what is driving these breaches of planetary boundaries. Most pertinently, as far as the future of this planet is concerned – and bearing in mind that we live 'on a planet with intricately structured ecosystems and a delicately balanced climate' – Raworth asks this crucially-important question: 'how big can the global economy's throughflow of matter and energy be in relation to the biosphere before it disrupts the very planetary life-support systems on which our well-being depends? The evidence which has emerged since the start of this century would suggest that it's already too big!'[4]

As Wilson commented in 2016: 'For the first time in history a conviction has developed among those who can actually think more than a decade ahead that we are playing a global endgame.' The reasons for this include growing shortages of reliable freshwater supplies, climate change, pollution of the atmosphere and the seas, and destruction of natural habitats. The result of all this is that: 'For many species it is already fatal.' As Wilson has argued, because these problems 'are global and progressive, because the prospect of a point of no return is fast approaching, the problems can't be solved piecemeal.'[5]

Wilson's radical conclusion was that we can only hope 'to save the immensity of life-forms' on Earth by 'committing half of the planet's surface to nature.' He then warned that 'Unless humanity learns a great deal more about global diversity and moves quickly to protect it, we will soon lose most of the species composing life on Earth.' His argument in *Half-Earth* was based on an earlier suggestion, first made in 2002 – and he saw his suggestion as 'a first, emergency solution commensurate with the magnitude of the problem', as he was convinced that 'only by setting aside half the planet in reserve, or more, can we save the living part of the environment and achieve the stabilization required for our own survival.'[6]

Mass extinctions

Since the eighteenth century, geologists, palaeontologists and biologists have been finding out species extinctions – and mass extinctions in particular. Essentially, Earth's history – and the history of the species living on it – has seen long periods of relative stability. Up to the Industrial Revolution – or, in other words, the birth of industrial capitalism – the Earth had experienced five mass extinctions. According to scientists, we are now living through the Sixth Mass Extinction, which is predicted to be the most devastating extinction since the asteroid impact that wiped out the dinosaurs and much else.

Extensive and painstaking research has established that, in previous geological epochs, there has always been an average 'background extinction rate'. For most of Earth's history, though, extinction very rarely took place, and varied from species to species. According to palaeontologists and biodiversity experts, the normal pre-human extinction rate of species was approximately one species per million species a year. But now 'it is believed that the current rate of extinction overall is between one hundred and one thousand times higher than it was originally.' The higher rate was confirmed in 2015 by a study conducted by a team of international researchers – and, 'All the available evidence points to the same two conclusions. First, the Sixth Extinction is under way; and second, human activity is its driving force.' This confirmed Wilson's conclusion that 'Every expansion of human activity reduces the population size of more and more species, raising their vulnerability and the rate of extinction accordingly.'[7]

Although the conservation movement has slowed the species extinction rate, it has 'failed to bring it anywhere close to the pre-human level. At the same time the birth rate of species is dropping rapidly.' It is now clear that 'due to habitat loss alone, the rate of extinction is rising in most parts of the world.' As well as the extinction of known species, there is also the question of how many currently-unknown species are also being destroyed – presumably at the same rate of extinction. If the destruction of Mother Earth is allowed to continue at this rate, we risk entering what Wilson called the Eremocene, or the Age of Loneliness: 'The Eremocene is basically the age of people, our domesticated plants and animals, and our croplands all around the world as far as the eye can see.'[8]

As regards global biodiversity, extinctions and extinction rates, it's important to note that human actions causing extinction are synergistic: in that, as one action intensifies, other actions also intensify, so that the combined changes accelerate the rate of extinction: 'Clearing a forest for agriculture reduces habitat, diminishes carbon capture, and introduces pollutants that are carried downstream to degrade otherwise pure acquatic habitats en route. With the disappearance of any predator or herbivore species, the remainder of the ecosystem is altered, sometimes catastrophically.'[9]

The five most ruinous of modern human activities, in order of importance, are: habitat destruction, for farming and via climate change; invasive species; pollution of freshwater ecosystems and the atmosphere; population growth; over-exploitation of species via intensive fishing and hunting. All of these, separately or in combination, lead to the decline of species' numbers – making them more vulnerable to stresses such as disease and competition. The result: extinction of those species.

As early as 1962, Rachel Carson, in *Silent Spring*, drew attention to the dangers posed by chemical pesticides – especially in relation to *insects* – to the importance of 'the problem of sharing our earth with other creatures.' This was similar to what Marx had called the 'metabolic relationship' between humans and the rest of nature. According to Carson, humans needed to be aware that life involves 'living populations and all their pressures and counter-pressures, their surges and recessions.' Her warning was clear, that only by 'taking account of such life forces and by cautiously seeking to guide them into channels favourable to ourselves can we hope to achieve a reasonable accommodation between the insect hordes and ourselves.' Of particular significance, Carson – echoing Engels's earlier conclusion – stated that ideas about 'controlling nature' were 'conceived in arrogance, born of the Neanderthal age of biology and philosophy, when it was supposed that nature exists for the convenience of man.[10]

As well as global boiling affecting human life on Earth via increasingly-disruptive and destructive extreme weather events – whether floods, droughts or deadly heatwaves – it is also an important factor in negatively-impacting Nature's ecosystems. This is potentially the most serious threat to the life of all Earthlings.

Ecosystems

The whole concept of 'ecosystem' – a crucial element in the ecological crisis – originated with Arthur Tansley who, as early as 1913, had helped found the British Ecological Society and who had become its first president. Tansley had studied under the leading biologist Ray Lankaster, who was a close friend and supporter of Karl Marx. From his studies of plants, Tansley – by the

1930s, already Britain's leading ecologist – noted that the natural world had already been 'transformed' in many ways by humans, so that for some considerable time, the conditions of nature were the result of both natural *and* human history. In 1935, he argued that an ecosystem was the 'whole system', comprising 'not only the organism-complex, but the whole complex of physical factors forming what we call the environment of the biome – the habitat factors in the widest sense.' Though specific organisms might be of initial interest, the fundamenatally-important thing was to realize that 'we cannot separate them from their special environment, with which they form one physical system.'[11]

As Tansley explained, these ecosystems are parts of larger ones, that overlap, interlock and interact with each other, with a reciprocal relation between organisms and the larger material environment. Significantly, he also noted how the climate impacted ecosystems, and argued that whole 'webs of life' adjusted to various 'complexes' of environmental factors. In particular, he warned how 'the destructive human activities of the modern world' disrupted such systems, and that humans were an 'exceptionally powerful biotic factor which increasingly upsets the equilibrium of pre-existing ecosystems and eventually destroys them.' As a result, he argued that humans were capable of 'catastrophic destruction' of the environment.[12]

An important element of Tansley's work on ecosystems was that he elaborated a *dynamic* view of nature which argued that while there was often a natural stasis or equilibrium in the living world – a steady 'climax state' – contingency, disruption and even degeneration were also possible. In 1939, he argued that, while mature ecosystems had relative equilibrium, 'Positions of equilibrium' are seldom if ever really 'stable.' On the contrary, they contain many elements of instability and are very vulnerable to apparently small changes in the factor-complex.[13]

Significantly, Tansley insisted that humans could thus be seriously-destructive forces in nature – to the point of undermining entire ecosystems. It was this dynamic ecological-materialist understanding of ecosystems which underpinned the work of later ecologists such as Rachel Carson and Barry Commoner in the 1960s and '70s, who further emphasized how human activities were causing disjunctures and crises in existing ecosystems. As previously noted, this problem had been recognised by Marx who early on, while pointing out that humans had transformed the natural world which had existed before human history began, had also insisted that were fundamental ecological constraints – what we now call planetary boundaries – on what humans could do to nature, of which humans were a part and on which human society as a whole ultimately depended. The scale of the planetary crisis now confronting us is ultimately down, not to 'human activity' per se, but to the global spread of the capitalist juggernaut that continually strives for evermore capital accumulation via ever-expanding extraction, production, consumption and waste.

As we are seeing much more clearly today, the supposed human 'control' over nature – past and present – has encroached on and increasingly transformed it. With early moves from hunting and gathering to agriculture, human societies increasingly destroyed and cleared natural habitats to 'create' land for crops, animals and houses. The problems created by this were at first relatively small-scale and local. However, as Tansley noted in 1939, such destructive activity had increasingly spread during recent centuries, at: 'an increasing rate, all over the face of the globe except where human life has not yet succeeded in supporting itself.'[14]

Extinction rates

For mammals – the best-studied group of Earthlings – the 'normal' background extinction rate has been calculated at roughly '0.25 per million species-years. With around 5,500 mammal species alive in the twenty-first century, this should mean roughly one mammal species going extinct every 700 years. However, occasionally, the relative stability of life on Earth has been punctuated by rapid and devastating changes: these are what are called mass extinction events. During these, 'a significant proportion of the world's biota in a geologically insignificant amount of time' go extinct, leading to what David Jablonski has described as rapid 'substantial biodiversity losses ... that are global in extent.' Michael Benton, a leading palaeontologist, has used the metaphor of 'the tree of life', to describe mass extinction events thus: 'During a mass extinction, vast swathes of the tree are cut short, as if attacked by crazed, axe-wielding madmen.' Or, as described by Chris Packham and Andrew Cohen, the wrecking-ball that is 'human-induced climate change wreaks havoc on the planet.' They further note that planet Earth has 'always walked a tightrope' and that with too much extinction, 'the rates of extinction would go unchecked and the complex web of life begins to crumble.'[15]

In 2014, Elizabeth Kolbert pointed out that amphibians – with an extinction rate roughly 45,000 times higher than the background rate – had become 'the world's most endangered class of animals.' Particularly concerning was that she calculated that 'extinction rates among many other groups are approaching amphibian levels.' According to her, 'one-third of all reef-building corals, a third of all fresh-water mollusks, a third of sharks and rays, a quarter of all mammals, a fifth of all reptiles, and a sixth of all birds were 'headed toward oblivion.' Furthermore, these losses were spread all over the globe.'[16]

In 2004, a study was published – using the species-area relationship – to calculate the extinction risk posed by what was then called global warming, assuming species stayed where they were. According to their best-case estimates, if warming were held to what the IPCC were then defining as a 'minimum', between 22 and 31 per cent of the species would be 'committed to extinction' by 2050. But, as Kolbert noted, if warming were to reach what then was considered a likely maximum – which has since been passed – 'by the middle of this century, between 38 and 52 per cent of the species would be fated to disappear.' However, even under the more optimistic scenario – with species moving to new areas – with only minimum warming, from 9 to 13 per cent of all species would go extinct by 2050. With maximum warming, the figures were between 21 and 32 per cent. 'Taking the average of the two scenarios, and looking at a mid-range warming projection, the group concluded that 24 per cent of all species would be headed toward extinction.'[17]

At the time, the BBC's Science section headlined these findings thus: 'Climate Change Could Drive a Million of the World's Species to Extinction.' While the *National Geographic*'s headline was: 'By 2050 Warming to Doom a Million Species'. Though this study was soon criticised – for being either too pessimistic or too optimistic – subsequent developments have, sadly, more than confirmed their earlier estimates. In November 2023, a UN study concluded that, at current levels of global boiling and habitat destruction, the number of species at risk of extinction had doubled since 2004, to 2 million. Particularly worrying – because of their role in pollinating crops – was the risk of mass insect extinctions.[18]

Unlike other human disturbances of the natural world – which were and are usually limited to specific areas – climate change affects everything, everywhere. According to Kolbert in 2014, it would be necessary to go back fifteen million years in Earth's history to find CO_2 levels – and thus temperatures – higher than current levels. Her conclusion was stark: 'It's quite possible that by the end of this century, CO_2 levels could reach a level not seen since… the Eocene, some fifty million years ago.' While it is possible many species of plants – assuming they still possess the features that allowed their ancestors to survive and even thrive in those warmer ancient times – may be able to adapt to today's global boiling, we just don't know. As Miles Silman, a forest ecologist, commented that, assuming evolution works in the way it usually does, 'then the extinction scenario – we don't call it extinction, we talk about it as "biotic attrition", a nice euphemism – well, it starts to look apocalyptic.'[19]

In general, extinctions result from a multiplicity of factors, and there is what has been called 'a dark synergy' between 'fragmentation and global warming, just as there is between global warming and ocean acidification, and between global warming and invasive species, and between invasive species and fragmentation.' While, according to Tom Lovejoy, a US biologist, 'in the face of climate change, even natural climatic climate change, human activity has created an obstacle course for the dispersal of biodiversity,' which could result in 'one of the greatest biotic crises of all time.'[20]

Today's 'Sixth Mass Extinction'

Mainly as a consequence of capitalism's continued existence, today's Mass Extinction is happening at a totally unprecedented rate. Currently, the extinction rate is at least 1,000 times above the 'normal' rate which had existed before the capitalist economic system began to take off and spread. In fact, as early as 1997, the UN's Environment Programme raised the fear that it was possible that, if 'business-as-usual' continued, the extinction rate could soon rise to 10,000 times the natural rate. In 2019, in the most comprehensive assessment of its kind, the Intergovernmental Science-Policy Platform on Biodiversity and Ecosystem Services (IPBES) warned about the dangerous decline of Nature, at 'rates unprecedented in human history' and that the rate of species extinction was 'accelerating.' Then it warned that 1 million species were threatened with extinction – which would have 'grave impacts on people around the world.' In particular, Robert Watson, the Chair of the IPBES, after pointing out how the health of ecosystems – on which we and all other species depend – *was* 'deteriorating more rapidly than ever', warned that we were now eroding 'the very foundations of our economies, livelihoods, food security, health and quality of life worldwide.' However, on the positive side, the Report stated that 'it is not too late to make a difference, but only if we start now at every level from local to global.'[21]

Since Carson, Wilson and Kolbert have written, evidence that we are indeed living through a 'Mass Extinction' event continues to multiply. In November 2021, twenty-three US species were 'delisted' from the Endangered Species Act, as many of these species had not been seen in decades. But the announcement was a sobering reminder that the climate crisis and habitat destruction are accelerating an extinction crisis that now threatens millions of species. In November 2023, the journal *Plos One* reported results that showed that, overall, 19 per cent of European species are now threatened with extinction. This was higher than calculations made earlier by the Intergovernmental Platform on Biodiversity and Ecosystem Services (IPBES), which in 2019 had estimated it as 10 per cent. However, the extinction risks were higher for plants (27 per cent) and invertebrates (24 per cent). The latter is particularly important as 90 per cent of all invertebrates are insects, which provide vital ecosystem services: pollinating crops, recycling nutrients into soils, and decomposing waste. As lead researcher Axel Hochkirch noted: 'Without insects, our planet will not be able to survive.' These numbers were then extrapolated to make a global estimate of total species at risk of extinction – this confirmed previously mentioned findings: that two million plant and animal species are currently threatened with extinction. The Plos study concluded that 'Maintaining and restoring sustainable land and water use practices is crucial to minimise future biodiversity declines.' The most significant driver of these current declines is the expansion of capitalism's industrialized agriculture, which continues to destroy natural habitats at an alarming rate. Other important factors are the overexploitation of natural resources, pollution, and residential and commercial development. As noted by the Plos study: 'We are losing biodiversity and nature's contributions to people at rates never before seen in human history.'[22]

While, in the past, relatively small local areas have experienced extinctions of several species as a result of aspects of the economic systems adopted by humans, today we are facing planetary-level extinctions because of one particular type of economic system: capitalism. As capitalism has extended its reach beyond local areas to regions and continents, and then to the entire globe, it has increasingly 'pushed environmental degradation to the planetary level, as habitat destruction decimates the living conditions of species and as ecosystems are radically transformed.'[23]

Essentially, capitalism is destroying the natural world and thus the future of human civilization – rather like a vampire sucking the life-blood from its victims – in a totally unprecedented way. As the capitalist elite and their corporations swell with ever-increasing profits and accumulated capital, the natural world is being constantly drained of life. Part of the problem is that, as Marx noted, capitalism alienates humans from the rest of the natural world – and also creates the 'myth' that human society is not dependent on Nature and its ecosystems. This is, in large part, why this current Mass Extinction of species is so different from previous ones: and, for as long as capitalism exists and destroys the natural world in its relentless pursuit of short-term profits and capital accumulation, this extinction will continue apace. The more that capitalism carries on destroying habitats – for example, destroying ever-increasing parts of rainforests and consuming more and more freshwater in the interests of mass industrialized animal agriculture – the less ecosystems are able to recover. As a consequence, environments are 'simplified' and biodiversity declines – and ecosystems become less and less resilient, and thus increasingly unable to provide 'services' to humans, such as purifying freshwater or mitigating floods.

Particularly serious is that, as we have only recently discovered, a whole range of species in a given area often depend on each other, and have established complex interactions – such as trees sharing water or providing protections against pests or diseases. Thus, habitat destruction, the resulting loss of some species and the wider weakening of complex ecosystems, all risk triggering what are known as 'cascading extinctions.'

It is now clear to many – including those who nonetheless argue against any action – that humans have had, and increasingly are having, significant negative impacts on the Earth's systems. As with so many other aspects, if we go back to Ancient Greece, we find that Theophrastus – one of Aristotle's students – observed in about 350BCE in *The Warm and The Cold* that the draining of wetlands by humans had affected 'the moderating effects of water and led to greater extremes of cold, while clearing woodlands for agriculture exposed the land to the Sun and resulted in a warmer climate' Some thinkers during the Renaissance period also speculated that local weather systems were affected by building cities, chopping down forests and carrying out irrigation of farmlands. As regards the latter, in the 1750s, David Hume – writing at the time when capitalist agriculture was rapidly expanding via the so-called Agricultural Revolution – argued that 'advanced cultivation of the land had led to a gradual change in climate.'[24]

The loss of species which is happening now, is being rapidly accelerated by global heating – and this is when the increase in the average global temperature today is 'only' slightly above 1.5°C. Yet, in 2012, the World Bank had reported that scientists were almost unanimously predicting that, if no significant policy changes were undertaken, the average global temperature would have risen by 4°C by the end of this century – possibly by as early as 2060. In fact, a study in 2018 forecast that such a rise in average global temperature would see up to half of animal and plant species becoming extinct by the end of the century. Then, as has been seen, a UN study in 2019 predicted that some one million species were facing extinction in the near future. Such a collapse in biodiversity would put human communities at risk, as a result of loss of food sources, pollution of fresh water systems and the oceans, and erosion of natural defences against extreme weather events. Yet, so far, most pledges made at the various COP meetings have yet to be fully implemented. Of particular concern is the fact that earlier predictions regarding the timing and intensity of the impacts of global heating have proved too optimistic. However, an increase in average global temperatures is not the only thing behind the Sixth Mass Extinction – if anything, it's much more down to habitat destruction by industrialized animal agriculture.

Welcome to the Anthropocene

Officially, geologically-speaking, we are currently living in the Holocene epoch of the Cenozoic Era, which began around 11,700 years ago, when the last continental glaciers began to retreat and a milder climate allowed the emergence of probably the highest number of different species in Earth's history. However, many climate scientists and ecologists wonder if it should now instead be called the Anthropocene, because of the growing impact of humans on the natural world. This is especially so because of the mounting impacts of the increasing expansion and spread of capitalism from the eighteenth century onwards.

In 2000, the Dutch Nobel Prize winner and chemist Paul Crutzen first argued that various dramatic developments had pushed the Earth into a new state, and coined the term 'Anthropocene' to describe the current state of the world – to mark a distinct break from the Holocene geological epoch. For him, this new term made clear that we were living in a 'human-dominated geological epoch.' According to him, by 2000, human activity had transformed up to one half of the Earth's land surface, while humans were also using more than 50 per cent of the world's freshwater supplies, and more than 30 per cent of ocean life. More importantly, Crutzen drew attention to how humans were now altering the composition of the world's atmosphere. This observation was confirmed by the 2021 documentary, *Breaking Boundaries*, which features Johan Rockström explaining how we have left the Holocene behind. For instance, the ever-increasing use of fossil fuels and deforestation since the Industrial Revolution has added almost 550 billion metric tonnes of carbon to the atmosphere. As a result, the concentration of CO_2 in the air had increased 'by forty per cent over the last two centuries, while the concentration of methane, an even more potent greenhouse gas, [had] more than doubled.' Crutzen thus concluded that the global climate was likely to 'depart significantly from natural behaviour for many millennia to come.'[25]

That conclusion was based on noting that fundamental aspects of the Earth system – such as atmospheric composition, climate and ecosystems – had now sharply departed from the stability of the Holocene, which had allowed human civilization to grow and flourish. By 2000, the concentration of CO_2 in the atmosphere was over 400 parts per million (ppm) – certainly higher than at any point in the last 800,000 years; according to Kolbert, it was probably 'higher than at any point in the last several million years.' And, if current trends continue, by 2050, CO_2 concentration is likely to be over 500ppm – roughly double the pre-industrial level of 280ppm. Kolbert calculated that such an increase would produce 'an average global temperature rise of between two and four degrees Celsius.' which, in turn, would 'trigger a variety of world-altering events, including the disappearance of most remaining glaciers, the inundation of low-lying islands and coastal cities, and the melting of the Arctic ice cap.' However, as we now know, even at 410ppm – as the data for 2023 showed – we have already reached 1.5°C above pre-industrial levels. Perhaps corporate bosses and politicians should be forced to watch the 2009 film 'The Age of Stupid' every day, until they finally get the message?[26]

Many geologists and climate scientists – including those involved with the Anthropocene Working Group – have continued to gather evidence to support the proposal to term our period the Anthropocene. According to them, since around 1950, human activities have had such massive and unprecedented impacts that Earth's regulatory systems are now being overwhelmed to such an extent that a new geological epoch is needed. However, despite years of mounting evidence, in August 2023, an international sub-committee of geologists decided against doing so – though it did so amongst claims of voting irregularities. Nonetheless, as seen from the most recent studies and reports, planet Earth has clearly moved beyond the relatively-stable environmental conditions associated with the Holocene – with four key indicators over a period of 30,000 years showing a dramatic change from around 1950.

At the beginning of March 2024, an article in *Episodes*, the journal of International Geoscience, argued that the record shows that, since 1950, there had been 'an abrupt and major Earth System change consistent with the establishment of a new chronostratigraphic unit.' As of now, the global mean surface temperature is 1.5°C above pre-industrial levels – the highest it's been in 125,000 years. While atmospheric CO_2 levels are 146 per cent of Holocene levels – the highest for the past 3 million years; the figure for methane concentration is 257 per cent higher than Holocene levels: this concentration level, like that for CO_2, is also the highest for at least 800,000 years, as are nitrous oxide levels. Also significant, as has been seen, have been the rates of species extinction. The most worrying aspects of these changes are the speed *and* the magnitude of the changes over the past seventy years: indicating an epoch-scale change, rather than just a geological event. On graphs, these changes all display the hockey-stick curve, and represent 'what has been called the "great acceleration" of population, consumption, industrialization, technical innovation and globalization.'[27]

In the twenty-first century, we are now only too-painfully

aware that entire ecosystems are in crisis because of 'anthropo-genic' (human) causes – hence the increasing acceptance of the term 'Anthropocene' as a way to describe the current geological epoch we've been living through since the start of the Industrial Revolution. Like many environmentalists, however, Wilson and Kolbert put the blame on 'human activity' in general, rather than on capitalism itself.[28]

No one denies that humans have had a long – and often contradictory and complex – impact on planet Earth. Although through almost all of human history people have left geologi-cal evidences of their actions on Earth, they have not previously utterly overwhelmed it – though, in small local areas, some im-pacts have been quite destructive. Even the start of the Industrial Revolution did not have overwhelming impacts on a global or planetary scale. However, since the end of the Second World War, certain trends have 'swung into overdrive.' As a group of geolo-gists argued recently, the Anthropocene is 'real, it's already made geology, and it won't go away. Best to acknowledge it, to help us cope with the consequences.'[29]

As Kolbert made clear, the problems associated with what should now be called global boiling are only part of the crises currently happening. Oceans and seas cover 70 per cent of Earth's surface, with an exchange of gases between the atmosphere and the oceans taking place, in which gases from the atmosphere are absorbed by the oceans, and gases dissolved in the oceans are released into the atmosphere. Under 'normal' states of equilibrium, roughly the same amounts are dissolved as are released. However, changing the at-mosphere's composition – for instance, by adding huge amounts of CO_2 – significantly disrupts that exchange. This makes the oceans' surface waters more acidic – as Kolbert points out, even a small difference 'represents a very large real-world change.' As a result, oceans are now 30 per cent more acidic than they were in 1800. With continued 'business-as-usual' as regards the burning of fossil fuels, by 2100 'the oceans will be 150 per cent more acidic that they were at the start of the industrial revolution.'[30]

Approximately 30 per cent of the CO_2 that humans have so far released into the atmosphere – around 150 billion metric tonnes – has been absorbed by oceans. But, as Kolbert points out, 'it's not only the scale of the transfer but also the speed that's significant.' While there have been several instances in Earth's distant geological past in which severe ocean acidification took place, none of these past events showed the rapid rates of CO_2 release being seen in the Anthropocene. The nearest event comparable to such unprecedented rates was the end-Permian extinction – down to a massive prolonged burst of volcanic activity. However, this seems to have released 'on an annual basis, less carbon than our cars and factories and power plants.'[31]

As Kolbert states, by burning fossil fuels, humans have been putting CO_2 back into the air that had been sequestered for tens – in most cases hundreds – of millions of years. As a result, 'In the process, we are running geologic history not only in reverse but at warp speed.' As early as 2009, a study on ocean acidification noted that: 'It is the rate of CO_2 release that makes the current great experiment so geologically unusual, and quite probably unprecedented in earth history.' According to that study, if current trends continued – which they have – the likely result would be to 'leave a legacy of the Anthropocene as one of the most notable, if not cataclysmic events in the history of our planet.'[32]

As noted earlier, the world-renowned biologist Edward Wilson had argued that as much as half of the surface of the Earth needs to be returned to nature if catastrophic climate and biodiversity changes were to be avoided. According to his book, *Half-Earth* (2016), this is vital if we are to stave off the mass extinction of species – including of humans. His conclusion, based on mounting evidence, was clear: *'Humanity's grasp on the planet is not strong. It is growing weaker.'*[33]

The Capitalocene?

Like Wilson and Kolbert, many environmentalists still tend to see today's ever-worsening problems as the result of human activities in general, or even of 'human nature'. As humans are seen as the root cause of these problems, many such environmentalists are thus happy to accept the term 'Anthropocene'. However, given that the crises of so many ecosystems have seriously deepened since the mid-twentieth century, ecosocialists are among those who now argue that 'Capitalocene' – highlighting the particular impacts of global capitalism – might be a more appropriate or accurate term to describe our epoch. What Wilson, and so many other concerned ecologists and biologists, fail to do is to distinguish between 'human activities' in general, and those which result from the very essence or 'being' of capitalism itself. The latter is a social and economic system that requires constant economic 'growth' via ever-increasing production and consumption, in order to achieve constant capital accumulation – *regardless* of the impact of all this on the biosphere as a whole.

Ecosocialists argue that the evidence has been clear for the past two centuries that, if we are to solve the ever-worsening climate and ecological crises, we need to recognize that it is one historically-specific social and economic formation – capitalism – rather than human societies in general, which is driving planet Earth to the brink of catastrophic collapse. As Foster and Burkett have argued, Marx believed that the 'intensifying ecological problem of capitalist society' resulted from 'the rift in the metabolism between human beings and nature (that is, the alienation of nature) that formed the very basis of capitalism's existence as a system.' This systemic problem was made worse by capitalism's constant drive to expand in order to accumulate even more capital. Whilst these resulting crises have been deepening ever since the birth of capitalism, it is important to bear in mind that capitalism has only existed for a small part of human history. Keeping that short time-frame in mind – and the fact that *different* forms of social and economic organization *are* possible – is what allows the justifiable hope that we *can* build a better and more sustainable world.[34]

However, before developing that case further, the next chapter will examine the increasing threat of pandemics, which is closely related to the ecological crisis.

THE THREAT OF PANDEMICS

The reduction of the total number of wild animals like birds and bats has implications for our exposure to disease. Why? Because these are 'reservoir host populations' for pathogens, and the fewer birds and bats there are, then pathogen concentration and mixing tends to be higher (for reasons of lowered genetic diversity and easier spread). This increases 'spillover risk' for zoonotic infections to humans.[1]
Jem Bendell

Having had a sobering look at both the climate and the ecological crises, the bad news is that there's yet more 'doom and gloom' that needs to be taken on board before we can start developing realistic moves to a better world. The focus of this chapter is on a third major crisis: the growing threat of deadly pandemics. In fact, COVID-19 – like the climate and ecological crises – is yet further evidence of 'abrupt changes in humanity's relation to the earth in the twenty-first century.' Changes which, like the wider climate and ecological crises, are very much linked to the standard 'business model' of global capitalism. Thus, in many ways, COVID-19 can be read as a wake-up call to the fact that 'preserving the ecological equilibrium of the planet and therefore an environment favourable to living species, including ours, is incompatible with the expansive and destructive logic of the capitalist system.' In addition, this chapter will examine how the way many governments attempted to deal with the COVID-19 pandemic revealed a eugenics-based willingness to sacrifice older people, the disabled, those with serious underlying health issues, and those most economically-disadvantaged.[2]

As seen in the Introduction, Engels observed – when commenting on how, unlike other animals, humans attempt to master, as opposed to merely use, their environments – there can often be unintended and unexpected consequences from so-called 'victories over nature', as nature often took revenge. As he went on to explain: 'Thus at every step we are reminded that we by no means rule over nature like a conqueror over a foreign people, like someone standing outside nature – but that we, with flesh, blood and brain, belong to nature, and exist in its midst.' Thus, in a way, the devastating COVID-19 pandemic, which continues to mutate and to kill people across the world, can be seen as Nature's equivalent of Clint Eastwood's character 'Walt Kowalski' – in the film *Gran Torino* (2008) – taking its revenge on humans for the massive damage being done to its ecosystems. As Walt says in the film: 'Ever noticed how you come across somebody once in a while that you shouldn't have fucked with? That's me.'[3]

Where did COVID-19 originate?
Let's begin by making it clear that there'll be no discussion here of the various conspiracy 'theories' which quickly emerged on far-right populist and fascist social media platforms – and which continue to fascinate some people to this day.

Right-wing conspiracy theories notwithstanding, the COVID-19 pandemic was caused by a pathogen crossing over from the natural world to humans. It was not 'a natural event that no one could have anticipated', it did not come out of nowhere, it was not unexpected, and it was not unpredictable. As the historian Kyle Harper wrote: 'the pandemic was a perfectly inevitable disaster...[that] was inescapable.' In the first instance, what would quickly become a global pandemic began in China in November 2019; by January 2020, COVID-19 had already reached the USA. As Dr. Michael Greger – an internationally-recognized expert on public health issues – has revealed, the first cases were traced to the Hua'nan Seafood Wholesale Market in Wuhan: the largest wholesale seafood market in central China. Most significantly for the origins of COVID, that market reportedly also sold seventy-five species of wild animals: 'more than 90 per cent of the samples that turned up positive for the virus were found in the section of the...seafood market that trafficked in exotic animals sold for food.'[4]

Earlier epidemics – such as the Ebola outbreak of 2013-16 – had proved relatively easy to contain, as they were less transmissible and often occurred in areas where there is less international air travel. The COVID-19 virus, however – like all coronaviruses – is a 'stealth' virus which causes mass infection, has easy and rapid transmission, has varying symptoms, and can even be passed on by 'carriers' who either remain totally asymptomatic or who only show symptoms a week or more *after* infection. All those factors have allowed the virus to spread rapidly and extremely widely.

In May 2021, the World Health Organization (WHO) released updated figures for the true total global death toll of COVID-19, based on estimating excess mortality. According to their figures, by the end of 2020, total deaths from COVID-19 were at least three million. By the beginning of August 2024, the global number of confirmed excess deaths was over five million. However, according to a 'comprehensive and rigorous' assessment model developed by *The Economist*, 'the total number of excess deaths' during the pandemic was 'two to four times higher than the reported number of confirmed deaths due to COVID-19.'[5]

The COVID-19 pandemic has proved to be the biggest health challenge the world has faced since the so-called 'Spanish' flu pandemic of 1918-20. That virus – which is estimated to have killed some fifty million people worldwide – has been thought for some time to have been a zoonotic virus which originally crossed over to humans from wild birds, or possibly to humans via pigs. In fact, COVID-19 is the fourth such zoonotic epidemic/pandemic to hit humans this century. In 2002, and again in 2004, there was SARS, which was 'the first deadly global outbreak known to be triggered by a coronavirus', and which killed about one in ten of those catching this particular respiratory disease. Then, in 2012, there was MERS – which went on to kill one in three. Most of these appear to have originated in bats; in the case of MERS, it spread to camels, and from them to humans. In addition, from 2013-16, there was Ebola – though this was a filovirus, whereas the other two were coronaviruses. At the end of 2019, a scientific paper concluded that: 'The SARS epidemic demonstrated that novel highly pathogenic viruses crossing the animal-human barrier remain a major threat to global health security.'[6]

One aspect that all four infections have in common is that they were all viruses that crossed over from wildlife species to humans, sometimes via intensively-farmed animals. A second feature of these recent infections is that they can all be linked to the climate and ecological crises which have got worse since the start of this century. In fact, scientists and researchers have known for some time that disturbance and destruction of natural habitats – such as the Amazon rainforests – is one of the principal drivers of the transfer of animal-borne infectious diseases from wild animals to humans. According to Kate Jones, Chair of Ecology and Biodiversity at University College London, such developments are resulting in an 'increasing and very significant threat to global health, security and economies.' In 2008, she was part of a research team that determined that *at least* 60 per cent of the 335 new diseases that emerged between 1960 and 2004 originated with non-human animals.[7]

Preparing the way

One of the reasons why the COVID-19 pandemic was so deadly was because, in many countries – including in the UK – there had been decades of budget cuts to health and social care services: mostly in the name of capitalism's neoliberal mantra of cutting social expenditure programmes in order to increase 'growth'. Thus, many states around the world – in both the global North and in the global South – were ill-equipped to cope with the scale of the pandemic. Even many of the wealthiest countries were woefully unprepared when it first hit. In fact, it soon became clear that the COVID-19 pandemic had 'lain bare the fragility of existing economic systems.' The irrational and chaotic nature of capitalism is shown by the fact that the wealthier nations did have more than enough resources to ensure health services were properly equipped for such an emergency. Yet, in such countries, cutting taxes for the wealthy and underfunding public services on the one hand, combined with serial privatization on the other, had produced seriously-weakened health services. This aspect alone should be enough to convince citizens that the whole way in which our societies are organized and run should be overhauled from top-to-bottom.[8]

However, poorer countries particularly lacked sufficient health resources because of earlier colonial exploitation, and the more recent 'structural adjustment programmes' forced on such countries by global *pro-capitalist* organizations like the International Monetary Fund (IMF), the World Bank, and the European Central Bank. In the case of poorer nations, and of several of the wealthier ones, the health and social care services proved inadequate as so many governments had accepted the neoliberal view that 'too much' social spending would have meant higher taxes on the wealthy corporations and individuals, which would have impeded 'growth' and the accumulation of capital.

One of the duties of any government is to protect its citizens – not just from enemy attacks, but also from major health crises, such as pandemics. Yet the reality was that, even in the world's wealthiest countries – such as the US and the UK – this was not done. In the UK, from 2010 onwards, the Tory-LibDem coalition – and then, from 2015, the Tories on their own – allowed vital stocks of personal protective equipment (PPE) to 'degrade and go out-of-date.' In the UK, successive Tory governments also ignored and then suppressed the findings of 'Exercise Cygnus' – a three-day simulation exercise carried out in October 2016 to assess how the country would cope with a hypothetical H2N2 'swan flu' pandemic. The stark picture which emerged was that the UK was woefully ill-equipped to handle such a major pandemic, because of insufficient intensive and critical care beds, ventilators, and PPE for health staff. This was because of an ideological – i.e., neoliberal – approach which favoured underfunding the NHS, in preparation for privatizing as much of it as the general public would tolerate to private 'health' corporations.[9]

One consequence of this 'market' approach to health was that, even before the pandemic hit, 'bed occupancy in hospitals was between 94 and 98 per cent – which meant they were effectively full-up.' At the same time, the NHS was seriously under-staffed; it was thus no surprise that, when the pandemic first hit, overworked and inadequately-protected health and social care staff figured highly amongst the early victims. By May 2021, it was reported that 1,561 NHS and social care workers had died from COVID-19. The then government – headed by Theresa May – decided not to publish the 'Cygnus' findings. This policy was continued by Boris Johnson, when he took over in July 2019. However, in March 2020, *The Telegraph* reported the fact that Exercise Cygnus had taken place; then, in May 2020, in the midst of the COVID-19 pandemic, *The Guardian* leaked details of the findings.[10]

The writing was on the wall
Although many were surprised by the outbreak of the deadly COVID-19 pandemic, we had been warned. As early as 2005, the Marxist theorist Mike Davis had referred to just such a risk in his book, *The Monster at Our Door*. In that book, he pointed out that a number of factors 'were creating the conditions for a perfect storm of deadly viral infection on a world scale.' Those factors were clearly spelled out: the increasing appropriation and destruction of nature by capitalist corporations; the intensive industrialized rearing of animals in

massive agribusiness complexes; the growth of 'mega-cities' (many with large slums areas), often sited close to such agribusiness complexes; and the huge lack of basic health services amongst the poor in the global South. More recently, several health experts, such as Greger, had also called attention to the increasing susceptibility of humans to zoonotic pathogens.[11]

Essentially, both climate change, and the destruction and degradation of natural habitats and the resultant biodiversity loss, have made humans more vulnerable to such viruses. For instance, the destruction of natural habitats leads to declining food sources for wild species – such as bats – which, in turn, forces then to range into new areas, closer to humans. In addition, as explained in the quotation at the head of this chapter, the lack of sufficient food sources renders such species weaker and therefore more susceptible to infections.

Michael Löwy (who, in 2005, stated that 'All the warning signs are red') and Ian Angus are amongst those who have recently pointed out that capitalism's productivist operations are 'bringing the planet's inhabitants a long list of ... calamities' – including the likelihood of increased risk of new infections and pandemics because of the growing convergence of ecological crises. The latter listed most of these – and specifically highlighted the increased risks of deadly infections: 'Global warming ... Species extinction ... Deforestation...*New diseases and plagues*. The list goes on. We face a planetary emergency.'[12]

The pandemics crisis is thus just one more example of the consequences of the huge destruction of the natural world which continues apace. As early as 2009, the World Social Forum – involving more than four hundred activists representing thirty-four countries – met in Belém in Brazil and agreed what became known as the Belém Declaration. This had been drawn up by ecosocialists Ian Angus, Joel Kovel and Michael Löwy (with assistance from Danielle Follett), following a decision by the Paris Ecosocialist Conference in 2007. In 1916, Rosa Luxemburg had issued a stark warning – during the horrors of the First World War – that two choices were facing humanity: 'Socialism or barbarism.' The Belém Declaration essentially updated her message to 'Humanity today faces a stark choice: ecosocialism or barbarism' – though it would have been more accurate to speak of 'capitalist barbarism.' This sentiment was graphically encapsulated at that meeting by Evo Morales, the president of Bolivia, who said: 'The world is suffering from a fever due to climate change, and the disease is the capitalist development model.' Of particular relevance to the whole question of new pandemics, the Belém Declaration specifically warned – ten years before the emergence of COVID-19 – that 'epidemics of malaria, cholera, and even deadlier diseases will hit the poorest and most vulnerable members of every society.'[13]

Pandemics and the destruction of nature

In 2014, Paul Burkett noted that, in recent years, it was becoming more and more obvious 'that the natural conditions of human life (not to speak of other species of life) are increasingly threatened even as – indeed, precisely because – capital continues to accumulate.' Although this was a general comment about what capitalism was doing, it clearly has real relevance for understanding how the destruction of the natural world creates conditions for pandemics. This, in turn, is closely connected to Marx's comments about how capitalism's commodification of nature was leading to the growing degradation of nature, and a consequent dangerous 'metabolic rift' or separation between humans and the natural world. The historian and environmentalist Andreas Malm saw Marx's concept of the metabolic rift as being a contribution to understanding environmental problems as one that 'has outshone all others in creativity and productivity.'[14]

The importance of Marx's concept for understanding the causes of ecological crises is based on him understanding that the worsening ecological problems resulting from capitalist operations were down to the metabolic rift those operations increasingly created between human beings and nature. The alienation of humans from the rest of nature – 'that formed the very basis of capitalism's existence as a system' – was made worse by the constant drive for yet more capital accumulation: in other words, by *'capitalism's own expansion.'* What has happened, especially since 2000, is that modern capitalism's developments – such as the on-going destruction of the natural world and the intensification of animal agriculture, along with the 'rapid export and re-export of agricultural commodities, as well as globalised air travel' – has accelerated the transmission of pathogens like the one that causes COVID-19.[15]

In 2005, Mike Davis warned that zoonotic diseases had become 'far more more common in the era of industrialised agriculture' – and that, because of globalization, they spread much more quickly. This was confirmed in March 2020, by an article in which three scientists made it clear that such lethal zoonotic pathogens as COVID-19 emerge from two main sources: intensive complexes of industrialized animal agriculture; and, especially and increasingly, from wild animals because of deforestation or because they are brought into meat markets in large cities as 'exotic foods.' Both are aspects the researchers see as arising 'out of the capitalist mode of production.' which they see as 'destroying regional environmental complexity that keeps virulent pathogen population growth in check.' Not just because of the cramped and over-crowded conditions in which around 80 per cent of all farmed animals are now kept, but also because neoliberalism has ensured that government inspection regimes are cut back, while powers to enforce proper standards are reduced. It is this which creates the conditions for 'neoliberal disease emergence.' But even more important is that continued capitalist deforestation is making it increasingly likely that pathogens will cross from the wild to humans because the drive for evermore capital accumulation is destroying the 'ecosystems in which such "wild" viruses were in part controlled by the complexities of the tropical forest.'[16]

By April 2020, the COVID-19 pandemic had become global, and was in full devastating swing. That month – on 22 April, to be precise – also saw the fiftieth anniversary of Earth Day. On 22 April 1970, 20 million people in the US – around ten per cent of the USA's total population – took to the streets and university campuses, in the first ever Earth Day, to protest against environmental degradation, and to demand a new way forward for Planet Earth. In the early 1970s, this environmental movement achieved some notable successes in the US: such as the setting up of the Environmental Protection Agency, and the establishment of the principle that 'the polluter pays.'

Unfortunately – for the planet and all the Earthlings who live on it – the late 1970s saw an aggressively-rampant neoliberal US capitalism increasingly push back against those gains. Since then, as neoliberal capitalism has spread its tentacles to encompass the world (Che Guevara often referred to the capitalist USA as the 'Yankee Octopus'), it has greatly expanded the global impacts of the capitalist system – making those earlier environmental problems much worse. As noted earlier, capitalism's rapid and on-going destruction of the natural world has resulted in an unprecedented combination of threats: the ever-worsening climate and ecological crises, as evidenced by global boiling; and huge losses of ecosystems and biodiversity. As a direct result of all of that, the risk of dangerous and deadly zoonotic pathogens and viruses – crossing over from the dwindling numbers of wild animal species to humans – continues to increase. As noted by Neil Faulkner, 'the Covid-19 pandemic is only one instance of an accelerating breakdown in the relationship between the human species and the natural world.'[17]

While epidemics have occurred in the pre-capitalist past, mounting evidence suggests that the COVID-19 pandemic is closely connected to the ecological disequilibrium – or metabolic rift – that capitalism's relentless and increasingly reckless pursuit of growth and profit has created, and which it continues to deepen. As noted by Kallis et al., the increasing ease with which viruses jump from animals to humans is largely the result of the 'expansion of corporate agricultural systems, encroachment of humans on habitats, and the commodification of wildlife, all integral to current growth economies.' An additional factor which helped the speed and extent of COVID-19's spread was the 'interconnectivities of accelerated global economies, … via airplane and ship routes.'[18]

Today's various mounting and interrelated crises have made it increasingly clear that capitalism's drive for unlimited and continuous production and consumption is just not ecologically sustainable. As early as 2005, Sheila Malone emphasized that: 'Capitalism operates on the basis that the earth's resources are there for limitless exploitation, and that market forces will always find a (benign) solution to a crisis.' The COVID-19 pandemic has shown that that is just another neoliberal myth.[19]

Eating our way to pandemics

However, it's important to note that it's not just global heating and climate change, and the consequent increasingly-frequent extreme weather events caused by fossil-fuelled capitalism, that is causing the loss of natural habitats and biodiversity. One of the biggest drivers of the destruction of natural habitats – and of the resulting 'Sixth Mass Extinction' of species – is the global capitalist agricultural system. This is especially true of the industrialized meat and dairy industries, which, firstly, destroy ever-larger sections of the natural world; and, secondly, also create unhealthy conditions for factory-farmed animals, which make it much easier for animal viruses to cross-over to humans. In addition, there is the use and abuse of wild animals – such as the capturing, breeding and eating of various species.

As Ian Angus commented in May 2024, 'If you wanted to build a pandemic-creation machine, you could hardly improve on the factory farm system' that has been created by modern capitalism. Or, as observed earlier by Rob Wallace, the development of big capitalist agri-businesses results in pratices that accelerate 'the evolution of pathogen virulence and subsequent transmission.' While the crowded conditions of industrially-farmed animals 'depress immune responses' and 'larger farm animal population sizes and densities of factory farms facilitate greater transmission and recurrent infection' The conclusion drawn by Ian Angus is clear: 'The accelerating emergence of zoonotic diseases is inextricably connected to the industrialization of poultry, pigs and cattle, which itself is inextricably bound up with capital's drive to expand, no matter what damage it does.'[20]

Yet the destructive aspects of the capitalist agri-businesses' drive for ever-greater production and profit was visibly apparent even during the COVID pandemic. While the world was reeling from this particular zoonotic virus, Bolsonaro stepped up the destruction of the tropical rainforest in Brazil to provide more land for the cattle industry – in just over six months, satellite photographs showed that an area the size of Germany had been cleared. Yet scientists and researchers have known for some time that disturbance and destruction of such natural habitats is one of the principal drivers of the transfer of animal-borne infectious diseases from wild animals to humans.

Dealing with the wider ecological dimensions of such pandemics will clearly involve, amongst other things, *a revolution in what we eat*. To deal with the wider ecological dimensions of this pandemic, as Alan Thornett explained in a very timely article in 2020, will involve 'a revolution in the infrastructure, [in] how we live; the size of cities, how we travel, and what we eat. The task is gigantic but there is no alternative if we are to forge a sustainable future for the planet which resolves the contradiction between ourselves as modern humans and [the] myriad of other non-human species we live alongside.'[21]

Years before COVID-19 burst into the world with such devastating and far-reaching effects, concerns had been raised about the implications for human health arising from capitalist animal agriculture. Not only are industrial farming practices cruel to animals, they have also been judged as 'damaging to human health and the environment.' The push for greater and greater profits has resulted in methods to raise animals as quickly as possible results in 'inhumane crowding conditions for the animals, [and] the routine use of antibiotics and growth hormones.' In addition, 'no-graze' industrialized animal complexes require high-energy crops such as soya and corn to be fed to cattle – 'even though they are capable of getting their entire nutritional requirements from pastures and hay.' In all, around 75 per cent of all soybean production is now used for animal feed.[22]

As well as being bad both for such intensively-farmed animals and for human health, the animal industry's need for soya and corn is one of the main drivers of the destruction of rainforests and other aspects of the natural world – such as the overuse and pollution of freshwater supplies. As pointed out by Thornett, in Brazil's Amazon alone 'the cattle sector … has been responsible for about 80% of all deforestation …, or roughly 14% of the world's total deforestation.' While, overall, modern capitalist agriculture 'uses 70 per cent of all available fresh water worldwide' and is directly responsible for 'sixty per cent of global biodiversity loss.'[23]

Modern animal agriculture is increasingly based on intensive livestock rearing in animal complexes and facilities. This was pioneered in the US, where such units became known as 'Concentrated Animal Feed Operations' (CAFOs), with each one containing over one thousand 'animal units', with confinement lasting for at least forty-five days a year. In 2006, the UN Food and Agriculture Organization (FAO) concluded that, as well as being responsible for 18 per cent of GHG emissions, animal agriculture also used – either directly or indirectly for the foodstuffs needed to feed them – 70 per cent of all arable land, and 30 per cent of the world's total land surface.[24]

The increased use of antibiotics in livestock farming is also bad news for those humans who consume meat and dairy products – as pointed out by Thornett, the WHO sees this as 'a major factor in the rise of superbugs and the increasing resistance of bacteria to the antibiotics available in our hospitals.' Such vast quantities of antibiotics are needed in order to control the infections which arise amongst these intensively-farmed animals as a result of being confined in cramped and crowded conditions. However, antibiotics are also used routinely on healthy animals to make sure none of the animals fall ill and thus reduce yields.[25]

As well as destroying great swathes of the natural world, the global capitalist agricultural system – especially the industrialized meat and dairy industries – thus also creates extremely unhealthy conditions for factory-farmed animals, which make it much easier for animal viruses to cross-over to different species – including to humans.

Disaster capitalism

Another aspect of the links between capitalism and the virus is how governments and business reacted. In many ways, the responses exemplified one aspect of what Naomi Klein has referred to as 'Disaster Capitalism': 'the idea of exploiting crisis and disaster' – and, as with Hurricane Katrina in 2005, the 'crisis exploitation' during the COVID-19 pandemic was also a crisis that capitalism had created in the first place, via its 'environmental, economic, and social framework.' As Klein said, 'with resource scarcity and climate change providing a steadily increasing flow of new disasters', neoliberal capitalism – and the governments that facilitate and serve it – now makes sure that all major 'relief and reconstruction' initiatives are contracted out to private companies: thus creating new 'market opportunities'. As the boundaries between governments and big corporations have become increasingly blurred under neoliberalism, a major feature has thus been 'huge transfers of public wealth to private hands, often accompanied by exploding debt, [and] an ever-widening chasm between the dazzling rich and the disposable poor.' [26]

In the UK, the initial response of the Johnson government – determined not to do anything that might impinge on corporate profits – was to adopt what has been described as the *'quasi-fascist notion of 'herd immunity'*: a form of eugenics which was prepared to 'sacrifice' hundreds of thousands of mainly older people, the poorest sections of society, or those with significant underlying health problems. Even when the mounting death toll forced a change in policy towards suppressing the virus, Johnson repeatedly and deliberately ignored medical experts and instead locked down too late and 'unlocked' too soon. The priority was clear – protecting the profits of private business took precedence over protecting the lives of British citizens. This was especially hard on those working in low-paid front line 'essential' jobs, who couldn't work from home, and whose homes were often too small to allow infected members to isolate from the rest of the household. This was why, by May

2020, Barrow-in-Furness (Cumbria) – with 'high levels of deprivation' – had an infection rate more than three times the English average. In addition, public bodies such as the NHS, local government and communities – often keen to implement possible actions and solutions – were bypassed, 'in order to hand lucrative contracts to private corporations for the provision of PPE, carrying out of tests, and the establishment of a track-and-trace system.' Few were surprised when it emerged – predictably – that the result of almost all these hand-outs to private enterprise was 'serial failure.' Other governments which also put business before lives as the pandemic hit – such as Trump's USA and Bolsonaro's Brazil – also ended up having similar dire results as regards deaths.[27]

Even the start of this devastating pandemic failed to weaken the adherence of neoliberal governments to maintaining the 'growth' desired by capitalist corporations. This in large part explains why some political 'leaders' were slow to take the necessary public health measures to protect people, and why such 'leaders' – such as Johnson – ignored scientific and medical advice, and instead tried to restart economic activity before it was safe to do so. Just how dangerous capitalist growth is for human health is underlined by the fact that some governments – in order not to impede the economic activities of capitalist corporations – had actually ignored scientific warnings, and instead even 'cut funding for pandemic research units and epidemic control teams.'[28]

Not 'if', but 'when'

It is now abundantly clear that a whole range of infectious diseases that have hit humans in recent years share a common 'origin story': human interaction with animals. Several experts now argue that human society must take radical steps to reduce the likelihood of even more devastating pandemics in the near future. According to Greger: 'The current coronavirus pandemic may be just a dress rehearsal for the coming plague…Consider a pandemic a hundred times worse than COVID-19, one with a fatality rate not of one in two hundred but rather a coin flip of one in two… With pandemics, it's never a matter of if, but when. A universal outbreak with more than a few per cent mortality wouldn't just threaten financial markets but civilization itself as we know it.'[29]

In many scientific and medical circles, a devastating pandemic – such as COVID-19 proved to be – was not unexpected. Back in

February 2018, the Director of the UN's WHO had warned that a pandemic could happen *at any time*, and that the world was not prepared for such a pandemic. Calling it 'Disease X', he felt such a pandemic was probable rather than possible. Mark Woolhouse, Professor of Infectious Disease Epidemiology at Edinburgh University, and some of his colleagues, had been behind the push to get the WHO to add 'Disease X' to the list of priority diseases. In 2021, Woolhouse predicted that another pandemic like COVID-19 could be 'around the corner' and that 'it is a matter of when, not if' that happens. This prediction confirmed what other researchers had warned the previous year: that *unless* 'we fundamentally adjust the modes by which we appropriate nature ... we are likely facing another deadly pandemic in far shorter time than the hundred-year lull since 1918.'[30]

At present, many medical scientists are watching the developing situation in the US as regards a highly-contagious sub-type of bird flu (H5N1) – which had originally crossed from poultry to wild birds and then to cows, and then to humans. At present, over 20 US states have recorded the presence of this virus. As yet, there is no evidence of human-to-human transmission. But some scientists outside the US see this as having 'true pandemic potential' if it mutates. The crucial statistic for humans is the Case Fatality Rate (CFR): the ratio of the number of confirmed human deaths resulting from infection to the number of those confirmed cases of human infection. So far, there have been only a handful deaths in the US. But earlier, in other countries – such as China, Cambodia, Thailand, Vietnam and Indonesia – it has shown an extremely deadly mortality rate of approximately 50 per cent. This compares to a mortality rate for COVID-19 of between 2 per cent to less than 20 per cent, depending on age group. In April 2024, India Bourke reported for the BBC that the first death from a H5N2 subtype of the virus had been reported in Mexico. According to Gregorio Torres, of the World Organization for Animal Health, 'scientists cannot yet predict if bird flu will become the next global human pandemic.' But the signs indicate that this particular disease is here to stay, thus the world needs to be prepared. According to him, 'every time there's a jump between species, it's a signal of potential increased risk.'[31]

As well as dangerous zoonotic viruses like COVID-19 and bird flu, there are also increasing problems with other and older illnesses. In April 2024, it was reported that, because of global boiling, dengue fever and malaria infections were now significant risks in Europe. Since 2023, the Asian tiger mosquito – which carries dengue fever – has become established in 13 European countries: Italy, France, Spain, Malta, Monaco, San Marino, Gibraltar, Liechtenstein, Switzerland, Germany, Austria, Greece and Portugal.[32]

Thus, ironically, the continuing destruction of the natural world – which is overwhelmingly the result of the capitalist system – could thus threaten the operation of the global capitalist economic system itself which, currently, is dominating the world, and which is the main driver of the ecological crisis. Old attitudes – shared both by capitalists, and by earlier socialists – about the 'conquest of nature' clearly need to be replaced by 'a radical conception of the need to restore the human social metabolism with nature while promoting genuine human equality.' The conclusion drawn by Faulkner – among others – is that neoliberal capitalism's economic and political system 'is increasingly incapable of upholding the life, liberty and well-being of the overwhelming majority.'[33]

A turning point?

If nothing else, the COVID-19 pandemic crisis made it painfully clear that 'System Change' is now needed as quickly as possible, in order to create an economic system that allows for a habitable and sustainable planet. Since that first Earth Day in 1970, the past half-century has shown that capitalism's imperative to continually push for ever-increased productivity and consumption, in order to expand short-term profitability and capital accumulation, is increasingly exposing the planet's ecosystems, natural habitats and species – including humans – to serious threats that are already significantly undermining the planet's ecological balance.

As deadly and disastrous as the COVID-19 pandemic has been, it has the potential to be a turning point. Although belatedly, the UK government did, after all find that – despite Theresa May's comments to the contrary in June 2019 – there was indeed a 'magic money tree'. While many scientific and health experts began pointing out that the effects of the pandemic showed that 'caring for people's health and wellbeing' actually made economic sense. The impact of the virus on different social groups also exposed – in a deadly way – just what deep economic inequalities had resulted from forty years of neoliberal economics and austerity. The same kinds of inequalities were also evident across the globe, showing –

once again – that those 'in more vulnerable identities and positions are likely to suffer more than others.' There were, however, also lots of examples of 'caring and community endeavors', despite the contagious and deadly nature of the virus. All the examples of mutual community aid which proliferated during the worst of the pandemic – and the willingness with which the vast majority were prepared to co-operate in social health measures in order to protect the most vulnerable – gave the lie to neoliberal capitalist propaganda that humans are, by nature, selfish and greedy. It is just one more reason why it *is* realistic to have the justifiable dream and the hope that, if we can but dispense with capitalism, we can indeed create a much better – and safer – world.[34]

By now, if you've read this far, it should be becoming clearer that the capitalist system is increasingly imposing a range of disasters on all Earth's inhabitants. As noted by Michael Löwy in 2015, all the warning signs are now 'red', and that it was increasingly clear that 'the insatiable quest for profits, the productivist and mercantilist logic of capitalist/industrial civilization is leading us into an ecological disaster of incalculable proportions.' Ultimately, capitalism's addiction to infinite 'growth' and expansion 'is threatening the natural foundations of human life on the planet.[35]

As has been seen previously, capitalist corporations have strongly resisted global attempts to significantly reduce greenhouse gas emissions, and to limit some of their other obviously-destructive actions – even though the negative human and environmental impacts of their actions are becoming clearer all the time. However, the COVID pandemic has shown that the capitalist system cannot even protect the health of the world's citizens. Essentially, capitalism is incapable of overcoming the planetary and ecological crises its operations have already triggered. Yet, despite increasing evidence of its destructive impacts on the finite natural world, capitalism remains fixed on its unsustainable goal of 'infinite growth.' Which is why, in 2016, Ian Angus concluded that: 'Capitalism has driven the Earth System to a crisis point in the relationship between humanity and the rest of nature. If business-as-usual continues, the first full century of the Anthropocene will be marked by rapid deterioration of our physical, social, and economic environment.'[36]

Sadly, we're not done with the 'doom scroll'. The next chapter will focus on another twenty-first century danger, also closely-related to the operations of global capitalism: the erosion of democracy.

Because, although the threat of future pandemics so clearly arises from the climate and ecological crises, the recent COVID-19 pandemic has also had very worrying impacts on politics around the world. Just as the current threat from new pandemics is very much connected with how capitalism has developed since the late 1970s, so too is what has been termed the 'hollowing out' of democracy.

THE TWILIGHT OF DEMOCRACY

The austerity experts' overarching goal was to bulwark economic relations from the influences of politics and state intervention; in doing so they both exempted economic policies from democratic decision-making and cast economic theory as apolitical.[1]
Clara Mattei

Though perhaps not immediately obvious, there are in fact several ways in which the growing threat of pandemics can be seen as a 'bridge' between, on the one hand, the climate and ecological crises and, on the other, the political – or democracy – crisis. The lack of real democratic control over the capitalist corporations which are driving the climate and ecological crises is one of the main reasons why zoonotic viruses are becoming an increasing threat. While, as we saw during the pandemic itself, right-wing populist and far-right groups used the pandemic to 'grow' their influence via conspiracy 'theories', and their various anti-lockdown, anti-vaccination, and anti-mask protests. The spread of their ideas on various forms of social media and on the streets – increasingly taking the form of 'creeping fascism' – is playing a significant role in the further weakening and undermining of democracy.

The democracy crisis we're currently facing – sometimes referred to as the 'Twilight of Democracy' – is composed of two main interlinked factors: rising social and economic inequalities, and the 'hollowing-out' of democracy: both of which are essentially the result of the operations of neoliberal capitalism. The fact is that today's neoliberal version of capitalism is as fundamentally opposed to democracy as the earliest forms of capitalism ever were. And, often, it's every bit as violent and destructive.

Capitalism and democracy

In many ways, to think of capitalism and democracy as going hand-in-hand is somewhat strange – if history is clearly looked at. Capitalism's birth was, as far as the vast majority of people were concerned, anything but democratic; while early capitalists fought tooth-and-nail to prevent any democratic control over their drive for profits, and to deny democratic rights for workers. In nineteenth-century Britain, where industrial capitalism first emerged, the various aspects associated with democracy – the right to vote, free speech and free assembly – were fiercely resisted by the capitalist class and their state. As noted by Neil Faulkner: 'Capitalist democracy has existed for a much shorter time than capitalism itself.'[2]

The lack of real democracy today is something that Extinction Rebellion UK has recognized for some time, with its longstanding call for Citizens' Assemblies. Most recently, from 30 August to 1 September 2024, they staged an 'Upgrade Democracy' event in Windsor. This made the point that the reason we've not had *meaningful* action on climate change and climate justice is because our democracies – especially since the mid-1970s – have been weakened, 'hollowed-out' and even 'captured' by capitalist corporations and their neoliberal ideology. As Julia Steinberger has argued: 'The horrifying fact is that we are not citizens in democracies, addressing our governments to redress injustice. If that were the case, we would have won a long time ago. We are disenfranchised people, facing governments taken over by industry actors, who welcome inequality and suffering as part of their ideology and business models.'[3]

For those who still buy into the myth that capitalism and democracy have always gone together like the proverbial 'horse and carriage' – at least for people living in the affluent countries of the global North – such negative views as those expressed above, about the state of democracy in the twenty-first century, may come as something as a shock. Yet when capitalism first really began to get established as the dominant economic system, genuine democracy was something the early capitalists fought long and hard against. In the UK, capitalists violently resisted the right for 'their' employees to form trade unions – and even to vote in elections – well into the late nineteenth century.

While, in the twentieth century, capitalism's economic crises after the First World War – including the Great Depression – led to the emergence of various authoritarian right-wing and fascist governments throughout Europe, which via severe repression imposed harsh economic measures on the many, to ensure the survival of... capitalism. Although it's true that, in most western European states after the Second World War, liberal democracy – accompanied by varying degrees of welfare capitalism – did flourish (in the UK, it came to be known as 'Butskellism'), it did so only in the context of relative economic prosperity, improved social welfare services, and rising living standards.

Significantly, when from the mid-1970s onwards, capitalism as a whole began to experience yet another economic crisis, the capitalist class – under the guise of what was then called 'monetarism' or 'supply-side economics', and which later came to be termed neoliberalism – began to erode all those social aspects and, along with them, increasingly mounted attacks on democratic rights. The first democratic rights targeted by early neoliberal regimes like those of Thatcher and Reagan were, of course, trade union rights.

Von Hayek's curse
The lack of real, effective, democracy for the many is reflected in the fact that – for the past forty years – it hasn't really mattered overmuch which mainstream party wins elections. Regardless of which party wins, the fundamental state of affairs is never really altered – because all mainstream parties are, to a greater or lesser extent, committed to neoliberal ideology and 'free market' capitalism. That neoliberal ideology is closely associated with Friedrich von Hayek and Ludwig von Mises. Ironically, although von Hayek's influential book was called *The Road to Serfdom* (1944) and was seemingly directed against the all-powerful states associated with both fascism and Stalinized communism, it was also a warning against what he called the 'tyranny' of the ideas of social democracy and the welfare capitalism that began to predominate in western democracies after the Second World War. Significantly, he started developing his ideas *in opposition* to the argument that pre-war fascism was a capitalist response to capitalist economic crisis and the challenges to it offered by socialism and communism.

The origins of neoliberalism can be traced to a conference in Paris in 1938, which von Hayek and von Mises both attended. It was at that meeting that the term 'neoliberalism' was first coined; and at that conference, they concluded that private property and competitive market economics were the only way to preserve 'freedom' from state interference: 'They believed that any form of collectivism – putting the interests of society before the individual – would lead inexorably to the kind of totalitarianism that had swept across Europe, in the form of Nazism and Communism. Except that by 'collectivism', they included the New Deal in the US, and the limited emergence of a welfare state in Britain – both of which were designed to curb the excesses of big business and to improve the lives of the majority.'[4]

In 1947, von Hayek formed the Mont Pèlerin Society (MPS) to promote his neoliberal argument – not surprisingly, those ideas quickly found favour with wealthy and powerful individuals, in both Europe and the US, who from the beginning had been opposed to developments such as progressive taxation, welfare states, and labour movements able to secure decent working conditions and wages for working people. Very quickly, the MPS developed into what has been termed a '"Neoliberal International" – a transatlantic network of academics, journalists and businesspeople seeking to develop a new way of seeing and running the world.' The MPS still exists to this day, to develop further ways in which the private sector can replace most of the functions still performed by governments. Naturally, it still claims to promote 'liberalism and free society.'[5]

It has been that very 'neoliberal' economic philosophy and practice – an 'invisible doctrine', as George Monbiot has described it – which has come to control the lives of us all, whether we've voted for it or not. This state of affairs really began from the late 1970s onwards, when neoliberalism's ideology and policies began super-charging the operations of global capitalism. And it is that economic doctrine – regardless of elections – which has played such a huge role in the environmental destruction and growing wealth inequalities which have spread plague-like across the globe.

Essentially, as summarized by Steinberger: 'Neoliberal ideology is antidemocratic at its very core. Its aim is to give free-reign over our societies to corporations, not citizens.' Hayek's neoliberalism was, in essence, a programme to destroy concepts of social justice (in his terms, 'distributive justice') and democracy for the many – as these would necessarily impinge on the absolute 'market freedom' desired by corporations. Taking neoliberalism to its logical conclusion, 'The imposition of market absolutism is a merely a means to an end: the goal is the destruction of democracy.'[6]

Despite banging on about 'freedom', neoliberal ideologues were actually intent on imposing an economic and political system which actually denies the majority of the population any real ability to plan their own lives or satisfy many of their social needs. It is no coincidence that the full force of neoliberal capitalism's war on democracy and social welfare was first fully put into practice in the brutal dictatorship established in Chile after Pinochet's US-backed military coup of 1973. The harsh neoliberal policies imposed under Pinochet's military dictatorship were largely drafted by the 'Chicago Boys' – economists associated with the University of Chicago where von Hayek had become a professor in the Committee on Social Thought in 1950 (significantly, his salary was funded by an outside foundation). The 'Chicago Boys' would go on to help impose similar policies in the eastern bloc countries, following the collapse of those one-party states. Their form of neoliberal 'economic shock therapy' was described by some of its victims as 'all shock and no therapy'. It is also no coincidence that one of Pinochet's most fervent admirers in the UK was Margaret Thatcher – who, from 1979, began imposing her brutal version of neoliberalism on the UK.

The Extreme Centre
Capitalism's 'capture' of mainstream parties – described by Tariq Ali as being slightly different versions of the 'Extreme Centre' – is today a global phenomenon. Long before Thatcher's Tory governments took this line in the UK after 1979, neoliberals had been preparing the ground. As early as 1955, an openly neoliberal think tank – the Institute for Economic Affairs (IEA) – was set up by Anthony Fisher, a businessman and a strong adherent of von Hayek's doctrines. It was the ideas of the IEA which 'captured' the far right of the Tory Party, and helped create a climate of ideas opposed to concepts of social welfare. The IEA also helped develop some of Thatcher's extreme neoliberal policies. Since then, Fisher has set up the Atlas Foundation – a network of over five hundred think tanks around the world – with the aim of promoting and strengthening neoliberal ideology in a strongly-co-ordinated and international way. Significantly, many of these think tanks have played a big part in promoting climate scepticism and denial – hardly a surprise, given that the funding of this network is not 'transparent', and that 'quite a bit of it, if not all, comes from the wealth of the extractive industries, first and foremost the fossil fuel industry.'[7]

In 1978, a senior Tory – concerned that a left-wing Labour Party might win the next election – noted that the UK's unwritten constitution contained the risk that parliamentary democracy might become 'an elective dictatorship'. However, since Thatcher led the Tories to victory in the 1979 election, it's arguable that our democratic processes have been eroded to the point where there is such a real 'democratic deficit' that we now have what amounts to a neoliberal 'elective dictatorship'.

Many are increasingly concerned about the state of democracy today. Tariq Ali, for one, has argued that because, in practice, 'all the mainstream parliamentary parties [are] acting in unison to defend capitalism', what we have – in most of Europe and the US, as well as in the UK – is effectively rule by the West's economic and political elites. These pro-capitalist neoliberal parties are what Ali calls the 'Extreme Centre.' According to him, as a consequence, 'democracy is in serious trouble especially in its European heartlands', with the result that the citizens of such states are in fact 'living through the twilight of democracy.' Whilst the West's elites had promoted their version of political democracy to tempt those living in Eastern Europe, they had simultaneously been 'quietly disencumbering themselves of that very system' – for the simple reason that global capitalism had 'no real need for a democratic structure, except as window dressing.' Today, in Hungary, Poland and Turkey – to name but three capitalist states – right-wing populist parties 'have cemented their position in power through the erosion of democratic rights and judicial independence' to turn their countries into 'illiberal democracies'. What should worry the citizens in western Europe is just how long the dominating neoliberal elites will continue to retain even the outward features of a democracy they have already almost totally drained of any meaningful content.[8]

In the UK, this hollowing-out of democracy stepped up a notch in the 2015 general election, when the Tories managed to get away with what can only be described as blatant election fraud via over-spending in 33 key marginal seats. After lengthy investigations, the Electoral Commission imposed a record £70,000 fine on the Tories for 'multiple offences.' The following year, the Crown Prosecution Service, although estimating that £118,124 was either 'incorrectly reported or not reported at all', decided not to prosecute in all but one of the cases under police investigation. However, its judgement was hardly a 'not guilty' verdict for the Tories: 'Although there is evidence to suggest the returns may have been inaccurate, there is insufficient evidence to prove to the criminal standard that any candidate or agent was dishonest. …. However, it is clear agents were told by Conservative Party headquarters that the costs were part of the national campaign.'[9]

In addition, during the EU Referendum campaign in 2016, the 'Leave' campaigns felt able to flout laws about donations; while in 2017, the Tories breached electoral law again – this time by the illegal use of call-centres to make 'cold calls' to specific voters in key marginal seats. In 2020, the long-delayed – and hugely-redacted – *Russia Report* showed the Tories had deliberately chosen *not* to investigate claims that Russia may have interfered in the 2016 referendum and even in the 2015 general election.

Significantly, Johnson and several other leading Tories had close links with various obscenely-wealthy Russian oligarchs who donated generously to Tory Party election funds. In December 2019, George Monbiot called attention to the way these oligarchs were *'gaming democracy'* – including playing a leading role in Corbyn's defeat that year – via lying, 'fake news', cheating, their control of the media, and misuse of social media, to persuade the poor to vote for parties that push the interests of the very rich; or to create a sense of powerlessness, thus leading to de-politicization. The following year, Monbiot highlighted research which showed that the UK is one of the world's most corrupt countries. Something that became even more apparent during the pandemic, with many contracts going to private companies owned by friends or even family members of government ministers and leading Tories.[10]

Equally concerning is how Johnson, almost as soon as he became PM in 2019, tried to act illegally, and to put pressure on the judiciary and senior civil servants. In August 2019, Johnson and his hard-right cabal lied to the head of state and prorogued parliament, to avoid parliamentary scrutiny of his Brexit agreement. This attack on UK democracy resulted in the Supreme Court unanimously ruling that his action was illegal because 'it had the effect of frustrating or preventing the ability of parliament to carry out its constitutional functions.' Ominously, Johnson very quickly announced plans to introduce significant changes to the judiciary – including to the Supreme Court. This was seen in many quarters as 'revenge' for the humiliating defeat it had inflicted on him. His bill to reform judicial review aimed to make whole areas of government action off-limits to the courts – including on immigration hearings. The Law Society warned that such moves threatened the legal curbs on state power. Johnson also announced plans to bring down a 'hard rain' on the Civil Service – seen by many as an attempt to move much closer to the US system of having political appointments in the top civil service posts.[11]

Since then, there's been the introduction of Voter Photo ID, which tends, in practice, to 'remove' large numbers of poor, young and ethnic minority voters from voting booths. In other words, those groups least likely to be Tory voters. According to the human rights organization Liberty, this move would damage election turnout, mostly affecting those who are already disadvantaged: 'If you're young, if you're a person of colour, if you're disabled, trans or you don't have a fixed address, you're much less likely to have photo ID and could therefore be shut off from voting.' Their judgement has been borne out by the Electoral Commission, which reported in September 2024 that almost 750,00 people did not vote in the July general election because of Voter Photo ID.[12]

Another part of the growing attack on democracy comes via 'Culture Wars', aspects of which often originated with hard-right individuals such as Steve Bannon and 'alt-right' groups in the US – targeting various groups, such as feminists, LGBTQ+ communities, trans people, migrants and refugees, and climate protesters. In the UK, it is trans people who are currently the main target of such 'Culture Wars' – as highlighted by Rowan Fortune when assessing the *Cass Report*, which was published in April 2024. Across Europe as a whole, governments and parties are increasingly adopting illiberal forms of governments and political agendas that borrow heavily from the authoritarian hard-right or even from fascism. However, the UK has its own 'Culture Wars' proponents – not just amongst the 'usual suspects' of the far right, such as Farage, but also on the Tory hard-right, with Suella Braverman and Kemi Badenoch being just two of the most extreme. Another worrying indication of the rightward drift of UK politics came in November 2024, when Badenoch won a large vote from Tory Party members, to become the new leader of the Tory Party.[13]

Rising inequalities

Despite all the 'promises' made by neoliberals, the greater 'freedom' they 'promised' has essentially been monopolized by the big corporations. One result of which is that they now pay very little tax, while in the UK ordinary citizens are taxed more heavily than at any time since 1948. Additionally, despite neoliberal rhetoric about reducing the 'big state', public spending hasn't fallen – as noted by Grace Blakeley, in her appropriately-titled book, *Vulture Capitalism*, it has simply been redistributed from the many into the off-shore bank accounts of the wealthiest members of society: 'Rather than spending on welfare and public services, deemed an inefficient use of public resources by neoliberals, states now spend billions on supporting big business and the wealthy with subsidies, tax breaks and bailouts.'[14]

As a result, apart from the climate, ecological and pandemic crises, another extremely clear – and, quite literally, also deadly – global aspect of neoliberal capitalism since the late 1970s has been ever-widening income and wealth inequalities. Such inequalities increased significantly following the 2008 global financial crisis – the most severe worldwide economic crisis since the Great Depression of the 1930s. Ever since 2010, an increasing number of people in the UK have experienced the painful impact of the austerity policies which have been imposed to 'solve' this crisis. However, as Clara Mattei has made clear, for some considerable length of time: 'austerity has been the mainstay of modern capitalism ...[and] a vital bulwark in defense of the capitalist system' – not just as a response to economic crises, but especially as a reaction to political crises when people, because of general social and economic dissatisfactions, begin to demand 'alternative forms of social organization.' And, as she also points out, there are many parallels between 'liberal' capitalism and fascism when it comes to dealing with such economic and political crises, when it is vital for the one per cent 'to rehabilitate capital accumulation.'[15]

The austerity weapon has in fact been used, since the 1920s, by various governments in states attempting to 'save' capitalism from its own chaotic economic and political operations. This has been done via various traditional methods: measures to restrict or even reduce real wages; reductions in the social wage, via cuts in public services; the privatization of public utilities, resulting in much higher costs to consumers; and taxation policies which essentially suck up money from the 99 per cent into the off-shore bank accounts of the one per cent.

The effects of such measures were shown by a UK study, carried out by University College London. This estimated in 2017 that austerity measures since 2010 had resulted in 120,000 'excess' deaths. This led Lawrence King, one of the contributors, to comment: 'It is not an exaggeration to call it economic murder.' One consequence of those austerity policies was a significant slowdown in what had been a decades-long trend of improvements in life expectancy in the UK, with women being particularly negatively impacted. This UK decline was higher than for any other leading industrial nation – apart from the US, which has long been seen as the 'capital' of neoliberalism. Since then, a follow-up study in 2022 reported that, up to 2019, there had been over 330,000 austerity-related deaths in the UK.[16]

Most recently, of course, people in the UK have been living through an unprecedented cost-of-living crisis. After a decade of declining real wages, April 2022 saw average wages fall by 4.5 per cent, the biggest fall since comparable records began in 2001. It was further predicted that the rate of inflation would top 18 per cent by the start of 2023 which, according to the Bank of England, would result in a very painful fall in real incomes over the next two years.

All this was happening at a time when big companies – whether the dirty energy companies or the 'privatized' utilities – were pushing up their prices whilst posting record profits and hand-outs to shareholders. Yet the neoliberal mantra from Sunak's Tory government was to continue urging workers not to push for wage rises that kept up with inflation. This was despite huge increases in home energy bills, and despite limited government interventions to reduce the full impact. The result has been an ever-increasing number of people becoming dependent not only on foodbanks but also now on so-called 'warm' or 'heat' banks. Back in 2010 – the first year of the Tory/LibDem neoliberal austerity attack on ordinary people – there'd been 66 foodbanks in the UK. In 2024, there are now around 2,000, with many of those using them being in work. According to the Trussell Trust – the UK's biggest network of foodbanks – the year ending March 2024 saw 'the highest ever levels of need as more people found their incomes did not cover the cost of essentials like heating and food.' During that year, they had distributed more than 3.1 million emergency parcels, with more than 1.1 million of them having been distributed for children. This was an increase of 94 per cent over the previous five years.[17]

Making the rise in energy bills even more disastrous, from both a living standard and a climate crisis perspective, has been government failures to deal with the UK's massive 'leaky homes' problem. Currently, the UK's housing stock needs retro-fitting, not just to save energy and thus GHG emissions, but also to help all those living in fuel poverty. The UK continues to be the worst in Europe for badly-insulated homes, and thus for the number of people dying of cold each winter – and that was before the newly-elected Labour government decided, in August 2024, to remove the winter fuel payment from most pensioners. Yet insulation programmes are virtually non-existent. Although the Tories had promised to increase public spending on this in 2019, the money never appeared.

The Climate Change Committee (CCC) has recommended that there must be increased funding for energy-efficiency, and that plans for this – especially for the poorest sections of society – need to be greatly accelerated. It thus makes massive sense – and is a vital aspect of social justice – that in this current cost of living crisis, climate and 'Net Zero' policies should be aligned to helping the poorest sections of our communities. Yet, instead, Tory governments up to 2024 still insisted on prioritizing hugely expensive and polluting dirty energy, while 1.5 million new-builds have failed to reach the standards needed for energy-efficiency.

As early as 2022, the Chief Executive of the CCC had commented thus on the government's record on insulating UK homes: 'It's a complete tale of woe.' It remains to be seen if the new Labour government's GB Energy scheme will do what's necessary. In addition, in May 2024, the *Financial Times* published a new OECD study which revealed that the UK has by far the highest rate of homelessness in the developed world, with over 50 people per 10,000 living on the streets or in temporary accommodation. It also revealed that homelessness was continuing to rise rapidly.[18]

Whilst the many have become poorer as a result of all this, the already-wealthy few have – unsurprisingly – become much wealthier. In April 2024, Forbes gave another indication of just how unequal things have become, by reporting that, in the previous year, another 141 people became billionaires: '26 more than the previous record, set in 2021.' Additionally, the world's 2,781 billionaires now own a staggering $14.2 trillion between them – '$2 trillion more than just a year ago and $1.1 trillion above the previous record, also set in 2021.' All this in just one year! Actually, just ten people accounted for $507billion of those gains over the past year. Meanwhile, the vast majority of people around the world have been trying to cope with 'cost of living' crises (or other forms of austerity) and, in many parts of the global South, with mass hunger.[19]

This what lay behind the call – made in late April 2024 – from Brazil, Germany, South Africa and Spain, for the world's billionaires to pay a minimum two per cent tax on 'their fast-growing wealth to raise £250bn a year for the global fight against poverty, inequality and global heating.' As they noted, that would be enough to cover the estimated cost of the damage caused by all the extreme weather events that hit the people of the world in 2023. It would also go some way to ensuring that everyone in society contributes 'to the common good in line with their ability to pay.' Right now, however, 'billionaires have the lowest effective tax rate of any social group [meaning that] people with the highest ability to pay tax [are] paying the least.' While in the UK, in October 2024, a coalition of leading economists, green campaigners, millionaires, fuel poverty groups and the trade union Unite called on the new Labour government to introduce a wealth tax in the Budget due at the end of that month. This coalition pointed out that the richest 250 families in the UK have a combined wealth of £748bn, with one per cent of the population owning more than the combined wealth of 70 per cent of the population. They also drew attention to the fact that the richest 0.1 per cent in the UK have carbon footprints twelve times that of the average person. Consequently, the coalition argued that such a wealth tax would both reduce the stark inequalities in [the UK] and help raise the vital funds needed to ensure that the transition to a greener, cleaner, more prosperous future is fair for everyone at home and abroad.' According to a separate Greenpeace report, a temporary National Renewal Tax of 2.5 per cent on all wealth above £10 million – affecting less than 75,000 people – would raise at least £130bn over the next 5 years.[20]

Hollowed-out democracy
Thus, since the late 1970s, economic inequalities have increased vastly with increasing numbers – even in the more affluent global North – facing a precarious existence: via unemployment, semi-employment, zero-hours contracts, 'cost of living' crises, mounting household debt, exorbitant rents or homelessness, dependence on foodbanks and warm/heat banks, along with declines in the availability and quality of many of those public and social services deemed 'necessary' in the years following the Second World War. At the same time, the super-rich have grabbed over 30 per cent of

all the wealth resulting from economic growth since the 1970s, whether via rising incomes or the increased values of their assets. Despite the COVID pandemic, by the end of 2020, the world's billionaires were worth $12 trillion.

Yet all 'Extreme Centre' parties, to a greater or lesser extent, continue to prioritize the interests of the one per cent. According to Ali, what we have under neoliberalism is 'the dictatorship of Capital', with politics – regardless of which mainstream 'Extreme Centre' party wins elections – now reduced to 'little more than concentrated economics', with the political state mainly functioning as 'the executive committee of financialized capitalism.' Essentially, the super-rich have been using the power of their corporations, and the corresponding weakness of democratic institutions, to hoard ever more wealth in their off-shore tax-free bank accounts. Currently, the five biggest corporations are earning more than a quarter of the world's population. This is being done at the same time that these corporations are simultaneously responsible for creating rising inequalities, trashing the planet, and ensuring there are no effective policies to address the increasing poverty, or the climate and ecological crises, that their greed is creating. Not surprisingly, such developments, and the impacts of these crises, also have significant political consequences for democracy.[21]

From the beginning, neoliberal capitalism, aided by the state, has targeted a wide range of democratic rights – initially dealing with the labour movement and its trade unions. Many people and organizations tended to stand aside whilst governments were attacking strikers and their unions – for instance, during the 1984-85 Miners' Strike in the UK. However, developments since then have increasing borne out the third verse of Niemöller's poem, *'First they came…'*:

> *Then they came for the trade unionists*
> *And I did not speak out*
> *Because I was not a trade unionist…*
> *Then they came for me*
> *And there was no one left to speak for me*

As a result of those early victories over trade unions, neoliberal governments have since moved to attack other rights – including those of peaceful protest and the right to speak freely when on trial, as Extinction Rebellion and Just Stop Oil climate protesters in the UK have found out. It was awareness of recent attacks on democratic rights that led Robert Kuttner to conclude, in the negative, concerning the question posed by the title of his 2018 book, *Can Democracy Survive Global Capitalism?* For many others too, it is increasingly clear that, in fact, 'democratic capitalism' in the twenty-first century is 'a contradiction in terms.' According to Kuttner, capitalism now can only get the high rate of profits it deems necessary by limiting workers' rights, dismantling or privatizing social welfare services, evading taxes and environmental controls, and undermining the economic security of nations. In particular, he sees the far right – and not the left – as most likely to benefit from the increasing political polarization which will result from the attacks of neoliberal capitalism on living standards, social welfare and democratic rights.[22]

As the twenty-first century develops, it's increasingly clear that the longer capitalism continues, the more democracy is undermined, in a variety of ways. As more and more people are thrust from the 'just about managing' position to having to make decisions about whether to 'heat or eat', and increasing numbers of workers are forced to turn to foodbanks in order to feed their families, we're moving into a 'crisis of legitimacy' situation for capitalism. This is made worse with each new revelation of corruption, record profits, and vastly increased wealth for the one per cent.

Determined not to make concessions on social welfare spending, capitalism instead turns to a much-favoured weapon to support its rule – the same one favoured by the Roman Empire, and all subsequent empires and undemocratic regimes: divide and rule. With so many peoples' lives becoming harder and worse, far right and fascist groups emerge which try to place the blame for all these ills on 'others' – especially if they can be easily identified by their skin colour, clothes, religion or lifestyles. This scapegoating is something the defenders of capitalism are only too happy to amplify: hence mainstream right-wing politicians – and even some social democratic parties – are increasingly prepared to echo many of the tropes pushed by the far right.

In 1989, with the fall of the various 'socialist' regimes in eastern Europe, Francis Fukuyama was one of the first to argue that 'democratic liberal' capitalism had triumphed around the world – terming the new period of history then opening up as the 'End of History'. As we now know, although most of those eastern bloc states certainly got capitalism – mostly via what was called 'economic shock therapy' – they didn't get what we could call liberal democracy. Putin's Russia is just one striking example; while Orbán's 'illiberal democracy' in Hungary is yet another.

An illiberal democracy

As opinion polls began to show increasing public awareness of the climate crisis, and as climate activist groups such as Extinction Rebellion, Just Stop Oil and Youth Demand stepped up their civil resistance protests, hard-right Tories became increasingly vocal in calling for even further limitations on certain political freedoms and civil liberties which the climate movement and trade unions enjoyed. The Tories thus significantly limited the right to protest using peaceful (non-violent direct action) tactics – the Police, Crime Sentencing Act (2022), and the Public Order Act (2023), have given governments and the police wide-ranging powers to suppress any protest they don't want. This was apparent, for instance, in Braverman's actions – whilst still Home Secretary – in relation to protests in solidarity with Palestinians and against the genocide in Gaza. As Simon Pearson concluded at the end of 2023, Sunak's Tory government had continued to push an authoritarian agenda that threatens long-standing civil liberties in the UK. Their policies targeting protests, free speech, and dissent should deeply concern all who care about human rights and democratic accountability.' More recently, the Walney Report – 'Protecting our Democracy from Coercion' – which was published in April 2024, suggested that even harsher restrictions on peaceful protest are on the way. Although Walney was billed as an 'independent adviser', it has since emerged he has links to fossil fuel and weapons firms – the very companies rightly being targeted by protesters! Then, in July 2024, four JSO protsters were jailed for four years, and one for five years, for planning to pracefully disrupt the M25. These were the longest sentences ever handed down for peaceful protest. As of November 2024, it looks like Starmer's government will not be repealing the Tories' anti-protest laws.[23]

There has also been increasing evidence of a much more hard-line policing response to peaceful civil resistance. In October 2019, the police used a Section 14 Order against protesters involved in XR's 'Autumn Rebellion', requiring protests to end immediately; this tactic was later ruled unlawful by the High Court. Following Johnson's election victory in December 2019, policing of climate protests became increasingly restrictive. In February 2020, there was an unprecedentedly aggressive police response to a Greenpeace action outside BP's London HQ. Since then, there have been 'robust' policing of Extinction Rebellion and Just Stop Oil protests. In 2021, the policing of one XR protest led a retired Met detective sergeant to describe the police response as 'an appalling example of policing', and to state that: 'Clearly they've got orders to use force on protesters that are trying to protest peacefully in the street.'[24]

At the same time, Tory governments attempted to curb political democracy even further by repealing legislation that prohibits companies from hiring agency staff during a strike; while the Strikes (Minimum Service Levels) Act of 2023 made strikes by some public sector workers essentially illegal – something the Tories have had in mind since at least 2010.

If none of this gives you any 'reasons to be worried' (to misquote Ian Dury yet again!) about the state of UK democracy, in October 2021, the *Guardian* listed seven areas of particular concern about Johnson's post-2019 'power-grab' – and made specific parallels with recent illiberal developments in Hungary and Poland. As well as highlighting the likely voter-suppression impacts of Voter Photo ID, controls on the powers of the judiciary to oversee executive powers, and moves to severely restrict peaceful protest, the article also mentioned plans to restrict the activities of whistleblowers – by widening the scope of the Official Secrets Act, even where such leaks would clearly be in the public interest. Even *The Sun* called those moves a 'licence for cover-up', and commented about the UK being 'in the grip of oppression' if those plans went ahead. Other concerns were raised over plans to restrict the scope of the elections' regulator; political interference in Ofcom, which is supposed to be independent (just the sort of thing Orbán has done in Hungary); and using public money to bolster Tory MPs – of the 45 areas chosen to receive some of a £1bn fund to 'level up' poorer towns and cities, 39 had a Tory MP, even though next to them were areas which were even poorer. This *Guardian* article concluded by speculating whether, if all those developments were happening in Orbán's Hungary, or in Poland, the BBC World Service might report as follows: 'Amid fuel and food shortages, the government has moved to cement its grip on power. It's taking action against the courts, shrinking their ability to hold the ruling party to account, curbing citizens' right to protest and imposing new rules that would gag whistleblowers and sharply restrict freedom of the press. It's also moving against election monitors while changing voting rules, which observers say will hurt beleaguered opposition groups.'[25]

'Zbellion'

While increasing awareness of the urgency of the climate and ecological crises could act as a catalyst for the formation of united climate coalitions which could then develop democratic ways 'to protect humanity from the ravages of both a savagely unjust economic system and a destabilized climate system', there are also significant political risks. As early as 2014, Naomi Klein warned that climate change could also be 'a catalyst for a range of very different and far less desirable forms of social, political, and economic transformation.'[26]

A particularly worrying indication of how the capitalist elites are seriously thinking society will develop is in relation to private militias. As early as 2012, the weapons corporation Raytheon was fully aware of how things might develop, believing that 'demand for its military products and services as security concerns may arise as results of droughts, floods, and storm events occur as a result of climate change.' While, in 2018, George Monbiot mentioned how the author Douglas Rushkoff had revealed that the super-rich were already planning their 'escape routes' for when there is societal breakdown as a result of the 'climate shit' really hitting the fans. As Monbiot explained, some of the world's richest people – whose 'greed and self-interest' have caused this climate crisis – are attempting to 'secure their survival against the indignant mob' angered by what the elites have caused. It appears that now 'their most pressing concern is to find a refuge from climate breakdown, and economic and societal collapse.' While some are preparing to escape to Alaska or New Zealand, others are looking to hiring sufficient security guards; while a company called Survival Condo has been turning former missile silos in Kansas into fortified bunkers: every one so far completed has sold. According to Kate Soper, some of the richest people are describing the expected rising oceans, social chaos and political breakdown – arising from the environmental breakdown caused by capitalism – as the 'Event'. In 2022, Rushkoff updated his initial findings via his book *Survival of the Richest,* pointing out that the one per cent were now working on the 'insulation equation', by trying to calculate whether they could earn enough money to insulate themselves from the crises their greed was creating. He concluded that: 'Never before have our society's most powerful players assumed that the primary impact of their own conquests would be to render the world itself unliveable for everyone else.'[27]

It isn't only the rich who are fearing political breakdown because of the climate crisis, and the growing awareness amongst the 99 per cent that democracy has been undermined. Several governments in even the richest developed states fear that as the climate crisis worsens and inequalities increase, they will be faced by increasing opposition – especially from younger generations. For instance, in 2018 the USA's Pentagon identified 'Generation Z' – those born from the mid – to late-1990s to the early twenty-first century – as posing threats to the system. The Pentagon saw growing inequalities, precarious employment, rising unemployment and the inability to get into the housing market as likely to make this group in particular increasingly disaffected. In addition, the plan foresaw that the worsening climate crisis would lead to large numbers of climate refugees desperate to seek shelter in 'safe' countries like the US. The resulting 'war plan' – called 'Zbellion' – saw such resistance and 'problems' as coming as early as 2025![28]

While, in the UK, the last Tory government attempted to widen the reach of the Prevent list by including belief in socialism, communism, and anti-fascism – among others – as 'extremism.' In addition, there has been the increased out-sourcing of policing – and even aspects of war – to private security firms. Increasingly, the security guards of such firms are dealing directly with protesters, while the police often stand by and observe, as was seen in some of the HS2 protests. Additionally, some public spaces in major towns and cities are now being 'privatized' – in part, to prevent protesters from getting near company offices. This is something Margaret Atwood – whose dystopian novels often indicate that she has a finger on the pulse of our times – depicts in books like *MaddAddam* and *The Heart Goes Last*. A comprehensive academic – and alarming – treatment of these developments is provided by William I Robinson in his latest book, *The Global Police State*.

Although the courts at present are still able to intervene – for example, in May 2024, the High Court ruled that Braverman's decisions to restrict the right of peaceful protest were unlawful –

there are no grounds for complacency. As we saw over the Tory government's plans to deport asylum-seekers to Rwanda – which the courts had also ruled were 'unlawful' – the government merely ignored the courts and passed a new law which declared that Rwanda, against all the evidence, was 'safe'.[29]

In the end, we can only create a better world through restoring and then increasing real democracy – in our communities, our political systems, and, crucially, in our economic systems. That will entail abandoning neoliberalism and thus no longer allowing 'nameless faceless mega corporations to make predatory & destructive decisions.' Such genuinely-democratic structures and processes will almost certainly include citizens' assemblies and worker-consumer co-operatives. By working and learning together to come to democratic decisions, we will be able to weaken 'the neoliberal hegemons and their power grip on our societies, including the capture and corruption of our states.' By respecting the perspectives and needs of all (including all vulnerable minorities), by 'fostering of a culture of care and reciprocal work', and by replacing the nonsense of 'alternative facts' spouted by neoliberal capitalism's supporters with 'the acknowledgement of scientific reality', we will be able to make 'decisions to bring us back within planetary boundaries, while protecting the most vulnerable from the harms already upon us.'[30]

However, before we can make a serious start on that road to a better world, we will need to confront the particular dangers posed by the growth of extreme right-wing authoritarian populism and outright fascism – or, as it has been called, 'creeping fascism'.

CREEPING FASCISM

The Far Right and creeping fascism ... [are] rooted in the capitalist crisis and the political polarisation that are eroding the foundations of 'moderate' centrist liberal-parliamentary rule across the world.[1]
Neil Faulkner, et al.

So far, four distinct – but linked – modern-day crises have been examined: the climate crisis; the ecological crisis; the risk of further pandemics; and the neoliberal crisis, which has 'hollowed-out' of democracy and sharply increased inequalities. Worryingly, there is yet another crisis – again linked to the others already mentioned – which today plagues the twenty-first century: the spread of 'creeping fascism' and its ideas.

With even centre-left parties – having been 'captured' by neoliberal ideology – largely ignoring the needs of their traditional electoral supporters, and instead prioritising the demands of the economic elites, it should come as no surprise that, once more, dangerous authoritarian movements have emerged in most of the countries suffering the ravages of unrestrained capitalism. Many of these groups are far-right populist movements – but, behind them, various openly-fascist groups have mushroomed. All of these groups are eager to exploit the anger and rage of those on the sharp end of neoliberal policies and the various capitalist crises, which have undermined security for so many individuals and communities. Trump's victory in the USA's November 2024 presidential election shows what can happen if even the more 'progressive' parties fail to offer real solutions for their problems. As Michael Roberts has shown, around 40 per cent of registered voters did not vote, while almost 20 million failed to even register. The result – despite three out of four voters *not* voting for Trump – is that he's now in a position to implement 'Project 2025', as drawn up by the right-wing think tank, the Heritage Foundation. The result also shows the dangers of not voting for the 'lesser evil' – as Rebecca Solnit has said: 'Voting isn't a Valentine – it's a chess move.'[2]

For those who think fascism – in whatever guise – was largely finished-off after the end of the Second World War, and that even the harsh neoliberal form of capitalism is a 'million miles away' from the brutality of the fascism of the 1930s, it might be useful to recall that, as was seen in Chile after Pinochet's military coup in 1973, twenty-first century 'neoliberal ideologues align perfectly with brutal far-right dictatorships.' But, today, it's not just brutal military dictatorships, or even outright fascist regimes, we should be concerned about. Also worrying is the rise of far-right populist leaders and movements in 'democratic' countries – in both Europe and the USA – which 'profit' from neoliberalism's myriad crises. Such groups are just one way for the one per cent to ensure their 'Extreme Centre' holds firm and continues to protect their global dominance. At the very least, such groups – often deliberately – sow hate and division amongst the 99 per cent. This helps explain why, in the UK, the July 2024 general election saw Farage's hard-right populist Reform UK gain five MPs and win over four million votes, with 14 per cent of the votes – putting them third, ahead of the LibDems.[3]

'First wave' fascism

Fascism initially arose from the conditions and crises – largely created by capitalism – that impacted Europe just before and after the First World War. This particular form of fascism, which was predominant in the interwar period, has been termed 'first wave' fascism, and it tended to prioritize mobilizing large numbers of uniformed thugs on the streets. Fascism, despite – even in its early stages – sometimes using 'left' and 'anti-elite' slogans, always tried to use the poverty and suffering caused by capitalism to garner support, and then invariably came to the defence of… capitalism. To do so, it has always been prepared to use whatever methods are deemed necessary – these usually included the destruction of democracy and the crushing of labour and left-wing movements.

Essentially, then, early twentieth-century fascism attempted to attract and recruit the victims of capitalist crisis and inequalities. The first sections of society that early fascism tried to attract were the lower middle classes. Traditionally, it has been members of that social class who tend to be the first to rally to fascism. In fact, this has broadly remained true up to the present. Analysis of the 2016 Leave vote in the UK showed it was lower-middle class voters in the south – no doubt avid readers of *The Daily Mail* or *The Daily Express* – rather than 'left-behind' working-class voters in the north, who tipped the scales for Brexit. Contrary to what many still believe, 'most of the more organised and class-conscious sections of the working class voted Remain.' Later, early fascist groups then attempted to attract sections of the most deprived and oppressed classes under capitalism.[4]

One of the most insightful commentators on fascism was Leon Trotsky, the Russian revolutionary who was exiled (and eventually assassinated) on Stalin's orders, and who repeatedly warned about the dangers posed by early fascist movements. He quickly grasped that fascism was 'nothing else but capitalist reaction ... based upon the destruction of parliamentarism', and that in order to destroy democracy in a time of crisis, 'fascism unites and arms the scattered masses. Out of human dust it organizes combat detachments.' Thus capitalism, via fascist movements, was attempting to co-opt, for its own defence, 'all the countless human beings whom finance capital itself has brought to desperation and frenzy.'[5]

Trotsky's understanding of the essential nature of fascism led him, early on, to advocate a 'United Front Against Fascism' between the various parties of the European labour movement – including Germany's Social Democratic Party and the German Communist Party (KPD). However, his warnings in the late 1920s and early 1930s about the dangers posed by interwar fascism were largely ignored by the leaderships of left parties and the trade union movements. At the end of 1931, he directed this unambiguous warning to members of the KPD: 'Should fascism come to power, it will ride over your skulls and spines like a terrific tank. Your salvation lies in merciless struggle. And only a fighting unity with the Social Democratic workers can bring victory. Make haste, worker-Communists, you have very little time left.'[6]

The history of the rise of 'first wave' fascism shows clearly that for fascism – in whatever form – to grow and get into power, 'something big has to happen that disrupts the "ordinary world."' In the early twentieth century, such a 'Big Disruption' had been provided by the post-war impacts of the First World War and then the Great Depression. As not enough people heeded the warnings of Trotsky, and others, about the dangers posed by fascism – to democracy and to the labour movement – the world instead had to fight the immensely-destructive Second World War. But fascism was by no means eliminated from the politics of post-war Europe. Nonetheless, immediately after the end of that war, fascism – in the UK and elsewhere – still tended to focus mainly on the uniforms and street mobilisations that had been the hallmark of fascism before the war. However, because of the welfare capitalism that emerged in most of the countries of Europe after 1945, these groups had little success. Mosley's pre-war British Union of Fascists (BUF) collapsed, and fascism remained very much a 'fringe' phenomenon.[7]

In the UK, another 'Big Disruption' didn't emerge until the mid-1970s. Following the onset of the 1973 economic crisis triggered by rising oil prices, unemployment in the UK almost doubled. Making use of this crisis, the fascist/neo-Nazi National Front (NF) began to grow, reaching its zenith in the late 1970s and early 1980s. From then onwards, it began to decline and splinter – one of those splinter groups being the British National Party (BNP). There are important lessons to be drawn from the reasons why this fascist group declined. In large part, this decline was because of the establishment of organizations opposed to racism and fascism – such as the Anti-Nazi League and the Rock Against Racism initiative. These, and other, organizations punctured the rise of such fascist groups via a twin-track approach: by organizing meetings around unemployment, housing, social deprivation, and racism; but also, by countering – and sometimes physically blocking – fascist marches. More recently, Stand Up to Racism, Unite Against Fascism, and HOPE Not Hate, have all been active against far-right and fascist groups like the BNP and the English Defence League. However, as regards the decline of both the NF and the BNP, it's also important to recognize that another factor was mainstream 'Extreme Centre' parties taking on board, to some extent, some of the racist tropes of those fascist groups – this was particularly marked in relation to the Conservative Party.

Fascism and capitalist crisis

It wasn't just Trotsky who saw fascism as 'the temporary alliance of the [capitalist] elite and the mob' during a major crisis for the capitalist system – so too have others, including Hannah Arendt. At such historical moments, fascism operates in various ways which – sometimes consciously, sometimes coincidentally – protect or further the interests of the capitalist elites and corporations. In the past, its main aim was often to take political power – for instance, the economic crises and insecurities which hit Germany after the First World War played a crucial role in propelling Hitler and the Nazis into power in 1933. But, at other times – and especially more recently – it focusses instead on dividing and demoralizing the main victims of capitalism: the 99 per cent. This is done by using racism, sexism, misogyny, ableism, and intolerance of all those sections of society deemed 'alien' or 'other'. Significantly, though, fascism also targets the organized labour movement and democracy itself.[8]

The twenty-first century has of course already experienced multiple crises: the 2008 banking crash, extreme and prolonged austerity, the COVID pandemic and now, increasingly, the impacts of the climate and ecological crises. These neoliberal crises, and the 'hollowing-out' and corrupting of democracy, have created the crisis conditions for the rise of today's far-right populist and fascist parties. Fundamentally, fascism – which is essentially a movement of despair and rage – is endemic to capitalism because periodic crises are an inevitable feature of its operations. Today, capitalist crises have reached pandemic proportions because modern capitalism – via global neoliberalism – has succeeded, even in 'democratic' capitalist states, in eroding most democratic constraints on its activities. This, as examined in the previous chapter, has vastly increased the inequalities and injustices inherent in capitalism.

Although the current crises may not seem as dramatic as the Great Depression in 1930s, and, instead, mostly seem to be 'slow-burners', they are nonetheless serious systemic crises. Significantly, the impacts of those crises fall mostly on those who are least well-off. This is true, not just of the economic crises, but also – and increasingly – of the climate crisis. The results of that crisis – via extreme weather events such as floods, heatwaves and wildfires, and disruptions of food production and consequent food shortages – are having the greatest impacts on the least affluent, who tend to live in houses without air-conditioning, and many of whom live in fuel poverty. For instance, according to the Office of National Statistics (ONS), the 2022 heatwave resulted in 4,507 deaths in the UK linked to heat. In fact, since 1988, there have been over 50,000 heat-related deaths, and more than 200,000 deaths related to cold. Most of these deaths were linked to those living in homes without air-conditioning or adequate insulation. While a study by researchers from the University of Bristol, published in May 2024, showed that, in some areas of England and Wales, more people died from hot or cold weather than from COVID-19. This bore out what the UK government's former Chief Scientific Adviser had said at COP26 in 2021: that 'the climate crisis was a far bigger problem than COVID-19' and 'would prove more fatal without immediate changes.'[9]

In particular, neoliberalism has specifically weakened social or civil-society institutions – such as aspects of the welfare state, and trade unions based on collectivism. Of course, neoliberal policies are deliberately designed to 'make the poor and middle class poorer, [and] also fragilise welfare safety nets.' By doing so, especially by running down the welfare state, neoliberalism makes it much harder for employees to resist pay cuts, inflation and the removal of rights and protections at work. The result at work is thus increased atomization, exploitation and stress-levels, and feelings of isolation and dissatisfaction outside the world of work – made worse by the impacts of austerity and the 'cost of living' crisis, both of which are political *choices* made in order to benefit the one per cent. This, in turn has created an increasing sense of disempowerment, leading to demoralization and a political void which modern fascism exploits, and within which it can spread. Even in democracies, 'neoliberal policies contribute to the rise, even triumph, of far-right fascist movements' by creating economic insecurity and increased stress. All of this then 'creates fertile ground for the growth of the easy false problem, false solution proposals of fascism.'[10]

The combined impact of these multiple crises is creating a growing 'crisis of legitimacy' for capitalism's continued rule – in exactly the same way as the Great Depression did in the 1930s. One way for the one per cent to overcome this, is to try to shift the blame on to minoritie, using them as scapegoats, to get the oppressed to 'punch down' on those even less fortunate than themselves. As a result, the impacts of predatory capitalism on the 99 per cent, whilst at times strengthening radical left movements (such as the brief rise of 'Corbynism' in the UK 2015-19), have also led to the rise of ultra-right populist groups and the far right – and to growing social divisions. Essentially, 'the economic and social crisis that creates the opportunity for the Far Right and the slide towards fascism, creates a polarisation.' The results of that polarization are clear to see, from

Europe to Americas – including the rise of far-right populists such as Trump and Farage. Even the emergence of the oligarchs and Putin in Russia can be explained as the logical outcome of the harsh neoliberal 'Economic Shock Therapy' policies imposed by the West after the fall of the Soviet Union in 1991.[11]

The far right and 'Culture Wars'
Consequently, the current crises have given rise to various political developments on right – affecting traditional 'extreme centre' parties wedded to neoliberal capitalism, as well as the populist hard right, and the various far right/fascist parties, with the latter increasingly active on various social media platforms. The exploitation – and deliberate widening – of such polarizations by the far right is precisely why modern-day fascism – or 'creeping fascism' – poses real dangers for democracy. According to Paul Stocker, both the Brexit vote in the UK in 2016, and Trump's election in the US in 2020 – as well as the growth of the far right in Europe, posed 'the biggest threat to the liberal democratic order since the Second World War.'[12]

As has already been seen, neoliberal think tanks continue to spread their fundamentalist 'free' market ideas; to weaken even the limited climate actions adopted by some countries; or to promote non-existent technologies such as carbon capture and storage (CCS). But they also play a part in the ideological 'culture wars' which are making political discourse so toxic – especially on social media – and which are used by far-right populists and fascists to increase their electoral support. For instance, *The Guardian* reported in 2018 that several right-wing US think tanks, including the Middle East Forum – which had been involved in the Brexit campaign – had also helped fund fascist 'Tommy Robinson' with his legal battles: 'A rightwing American thinktank that spent a five-figure sum on Tommy Robinson's legal defence has said it is aware of up to four other similar organisations bankrolling a high-profile campaign to release him.'[13]

As a consequence of these multiple crises, and the polarization pushed by the far right, several hard-right populist leaders and parties have come to power in a range of countries over the past two decades – such as Italy, Brazil, Hungary, Turkey, and the USA. In the latter, although Trump was defeated in 2020 – and despite his attempted coup in 2021, and subsequent criminal convictions – he ended up winning the November 2024 election. Quite simply, he didn't 'go away', because capitalism's crises haven't gone away.

For fascism to grow, in addition to crises, there also has to be 'a large, persistent and unmistakeable perceived threat, around which…latent fears and prejudices…can be mobilized.' These include migration in general but increasingly now the number of climate refugees from the global South; feminism; equal rights for members of LGBTQ+ communities; and 'Culture Wars' issues in general. As Mason argues, it is essential to fight the right's 'Culture Wars' propaganda, and so 'weave anti-fascism into all aspects of popular culture.' This is particularly important as regards misogyny, patriarchal restoration, racism, xenophobia, and homophobia and transphobia – all of which are core to modern fascism. According to Mason, even more than a socialist revolution, modern fascists 'fear that the millennial generation of women, LGBTQ+ people and, yes, workers will demand freedoms unheard of in the twentieth century.'[14]

In fact, various think tanks linked to the Atlas Network also often 'dabble in divisive culture war topics, on gender (equal rights for women, queer & trans people), race or migration, for instance.' They do this – as stated by arch far-right populist Steve Bannon – to 'flood the zone with shit.' By raising divisive topics, and helping to undermine reasoned and compassionate collective discussion, they aim to destroy the foundations for reasoned democratic decision-making. They also produce 'fake' reports and experts – all as a way of sowing doubt and undermining genuine public (as opposed to private) expertise. Julia Steinberger is just one of those – Jeremy Walker (*More Heat Than Life: The Tangled Roots of Ecology, Energy, and Economics*) is another – who sees all this as 'a core part of [neoliberalism's] war on democracy.'[15]

Modern fascism

Today's fascism – sometimes called 'second wave' fascism – is, to a large extent, different from the 'first wave' fascism, and tends not to prioritize mobilizing large numbers of uniformed thugs on the streets. Instead, we have suit-wearing elite members posing as 'anti-elites' to spread what are essentially fascist ideas far and wide into mainstream politics – often in the form of 'Culture Wars' politics directed against a wide range of 'minority' groups. But modern fascism is also in favour of more extreme forms of neoliberalism – Farage supported Brexit partly because it would allow employers to make a 'bonfire' of the legislation which gave a modicum of rights to employees.

Another important difference between 'first wave' and 'second wave' fascism is how modern far-right and fascist groups are using social media and online platforms – such as YouTube, Telegram and '*chans' – to exploit capitalism's systemic crises in order to reach a much wider audience. As a result, several of their key themes have been picked up, and are being spread, by right-wing populists, authoritarian 'conservatives' and the 'alt-right' – the latter best described as an ' incredibly loose set of ideologies held together by what they oppose: feminism, Islam, the Black Lives Matter movement, political correctness, a fuzzy idea they call "globalism", and establishment politics of both the left and the right.'As Faulkner observed: '"the shit of ages" (Marx's term for the reactionary cocktail of right-wing politics) is uploaded, digitised to flow freely through a leaky online sewage system.' This is why the current strength of today's far right and fascist groups can't be judged simply from the number of people they can muster for their street mobilizations, because in the twenty-first century, the real mobilization is taking place online: 'For now, fascism's strength is best judged through the salience of their ideas, which have spread rapidly via social media.'[16]

There has been much debate about what these new right-wing movements and groups are – and especially about the main features of twenty-first-century fascism. The late Neil Faulkner termed the modern form of fascism 'creeping fascism.' Essentially, this is a 'combination of fascist reaction and corporate power', marked by extreme right-wing populism, nationalism, racism, and authoritarianism. Modern 'creeping fascism' attempts to 'shift the blame for social decay onto scapegoats'; and, like 'first-wave' fascism did, tries to bring large numbers of those suffering from the fallout of capitalism's multiple crises into 'a counter-revolutionary mass movement to protect the [system] from the explosive potential inherent in society's accumulating discontents.' Although 'creeping fascism'– with some exceptions – doesn't manifest itself like the 'first wave' fascists via large para-military groups, it is nonetheless apparent in the growth of parties like the French Rassemblement National (ex-Front National), the Alternativ für Deutschland (AfD), and especially Meloni's Fratelli d'Italia (FdI). The latter is a fascist party in all but name; yet, in 2022, Meloni was elected as PM of Italy. All three parties are characterised by taking 'strong' positions on national-ism and immigration, but also by having a hard fascist 'core'. Their campaigns against migrants and Muslims are essentially modern fascism's version of Nazi campaigns against Jewish people in the interwar period. And, as with the Nazis, these parties also target other minority groups. In 2021, in an update on his 'creeping fascism' thesis, Faulkner commented that the authoritarian right, and creeping fascism more generally, was *clearly* 'on the march, project-ing an ethno-nationalist dystopia, spreading the shit of ages, the old reactionary cocktail of racism, misogyny, homophobia, hatred, and violence.' Sadly, as the past three years have shown, Faulkner's assessment was absolutely right.[17]

Other observers of modern fascism, such as Enzo Traverso, have discussed what they call 'postfascism'. According to Traverso, this new version of fascism is a form of 'anti-politics' which has arisen in Europe and the US. This is directly linked to the way neoliberalism has hollowed-out politics and democracy itself. Because of the ev-er-growing domination of the capitalist elites, 'in the last three dec-ades, the alternation of power between centre-left and centre-right governments has not meant any essential policy change.' Increas-ingly – as has been seen in Britain and in the rest of Europe – power has shifted from legislatures to executives. In the UK, for instance, there were Johnson's attempts in 2019 to prevent parliamentary dis-cussion of his 'oven-ready' Brexit deal. While in 2024, Sunak's Tory government simply ignored the Supreme Court, which had ruled its Rwanda plan for asylum seekers and refugees as 'unsafe', by push-ing through a law which said Rwanda was – despite all evidence to the contrary – a 'safe' destination. While, within the EU, there have been the operations of the 'Troika'. The Troika – initially formed in 2010 to take over the Greek economy – is made up of bureaucrats from the European Commission, the European Central Bank and the IMF. Since then, the Troika has 'managed' the economies of several other EU states – according to the requirements of global capitalism. According to Traverso, 'parliaments are…compelled to simply rati-fy laws that have already been decided by the executive. In such a context, it is inevitable that 'anti-politics' will grow.' In this situation, postfascism – by emphasising national identity and a reactionary na-tionalism, and simultaneously attempting to merge people and an authoritarian 'leader' – 'purports to fill the vacuum that has been left by a politics reduced to the impolitical.'[18]

However, whether we call the modern form of fascism 'creeping fascism' or 'postfascism', what both terms have in common is that they accept that there are important differences, as well as similarities, between today's fascist groups and those of the 1920s and 1930s. Although modern fascism tends to emphasize xenophobia rather than racism, and Islamophobia rather than antisemitism, it also retains many tropes of 'first wave' fascism.

Gateways to fascism

Currently, UK fascism is deeply fragmented in terms of groups, but its central themes remain largely the same. While exploiting the economic misery of white 'left-behind' groups, the core ideas are still based on a fanatical and emotional British nationalism, the total rejection of 'multi-culturalism', the repatriation of non-white and east European immigrants and refugees, hostility towards LGBTQ+ communities and trans people, eugenics-based references to disabled people as 'useless eaters', and continued opposition to the European Union. Two glaring current examples in the UK of how 'creeping fascist' ideas are being mainstreamed by right-wing populists are provided by Farage and Tice's Reform UK; and by the 'National Conservatism' project, which counts Suella Braverman as one of its key supporters.

Over the past ten years, the increasing impacts of the climate and ecological crises, rising economic inequalities, and the hollowing-out of democracy, has led to political polarization around the world. A clear and significant reactionary bloc has emerged in most countries and shown, via elections, that it is prepared to vote for the authoritarian right. Overall, this reactionary bloc increasingly supports 'nationalism, racism, sexism, homophobia, transphobia, militarism, police power, conspiracy theory, [and] climate-change denial.'[19]

The forms of this reactionary bloc include economic nationalism (such as the 'Take Back Control' slogan of the Leave/Brexit campaigns); Islamophobia and cultural racism; and more general racism directed against migrants, refugees and asylum seekers, and members of Roma communities. Also significant – though elements vary according to specific groups – are antisemitism; 'moral panics' about 'law and order'; the erosion of 'traditional values' (hence opposition to equal rights for women, the LGBTQ+ communities, and trans people); the promotion of various conspiracy theories (e.g., during the COVID pandemic); support for increased state authoritarianism and repression via extra police powers to limit and crush civil resistance; and a foreign policy that privileges 'national' interests over international agreements and co-operation.

Far-right populists like Farage have long pushed such reactionary themes. Before his Reform UK, Farage's UKIP and then his Brexit Party had already been mainstreaming such ideas. Ironically, of course, Farage – the 'anti-elite' millionaire who has been called Britain's 'Poundland Trump' – threw a public tantrum when he could no longer bank with the one used by the royal family! And, bizarrely, he has even depicted John Lennon's song 'Imagine' as an 'elite' attempt to undermine 'Britishness' by forcing British people to get rid of their 'art, and personal loves and standards…families, languages and diversity in habitat, custom and culture, … identities and of everything that makes [us] human.' You really couldn't make it up! While the 'National Conservatism' group was described in 2023 as a possible 'gateway to the far right.' Increasingly, it seems any moral and intellectual dividing lines between ultra-conservatives and right-wing populists on one hand, and the far right and fascism on the other, are fast disappearing.[20]

The importance of these 'cultural' aspects of modern fascism cannot be overstressed. Although it is true that there were high votes for Brexit in UK areas pushed into economic deprivation by neoliberalism, the so-called 'anti-elite' revolt of 2016 – expressed by votes for Leave – was at least as much cultural as it was economic. Eric Kaufmann is one who sees the growth of the populist hard-right and the far right in the twenty-first century – in the UK, elsewhere in Europe, and in the US – as a 'right-wing backlash [which is] mainly about the question of immigration and ethnic transformation.' In the UK, Farage and his party dubbed the July 2024 general election 'the immigration election.'[21]

The anger of poverty

However, whilst it's important to remember that the Leave vote in the UK's EU Referendum in 2016 was largely delivered by lower middle-class voters in the south – and not by working-class voters in the north – the deprivation factor should not be dismissed. As Darren McGarvey has made clear, the many victims of neoliberalism do indeed experience multiple deprivations. The impacts of these – along with justifiable feelings of being unheard or ignored by the political elites – as well as resulting in apathy and an 'opting-out' of the political process, also, and unsurprisingly, often lead to feelings of simmering grievance, resentment, anger and outright rage: 'In communities all over Britain where people experience multiple levels of deprivation in health, housing and education and are effectively politically excluded, anger is felt. And this anger is something we will all have to get used to, unless things change.[22]

As has been seen, it is such understandable anger that right-wing populist and far right groups attempt to exploit, by turning such justifiable frustration and anger on to 'other' target-groups who are often even more deprived and ignored than they themselves, but who are nonetheless are seen – or are portrayed – as doing better than themselves. As a consequence of these right-wing misrepresentations, many amongst the deprived see such groups as 'always being privileged' above their own group, and benefitting 'from a slew of …advantages'. This is made worse by how the situation of deprived people – often living in poverty and in run-down social conditions – is reported and discussed in the media, leading to 'a culture that leaves many people feeling excluded, isolated or misrepresented and, therefore, adversarial or apathetic.' Such feelings are compounded by local politicians failing to sort out community problems, often resulting in people in deprived areas concluding that they're living in 'a run-down place that's been largely forgotten by wider society.'[23]

As was seen during the Brexit campaign of 2016 and, more recently, in the summer 2024 riots (dubbed the 'Farage riots' by James O'Brien on LBC radio), the far right deliberately stir up racism and xenophobia – whether against east European migrants or against asylum seekers – in an attempt to recruit support for their political project, by exploiting the justifiable discontent caused by increasingly nineteenth-century levels of inequality and the savage cuts in social services. Although much of this toxic 'hate' mix was spread by Farage's various political creations, and by outright fascist groups, the UK's right-wing media has also continued to play its usual disreputable role.[24]

Combatting fascism

Then, as now, what's crucial in combatting fascism – including its various modern 'second wave' incarnations which have largely, but not completely, replaced 'first wave' fascism – is 'our ability to recognise fascism for what it is when it confronts us.' Neil Faulkner went further, concluding that: 'Indeed, the whole future of humanity hinges on our ability' to do so. Yet recent history has shown that, all too often, warning signs are still ignored – even by those groups which should have been expected to offer clear opposition. Although fascism in our century – apart from a few small fringe groups – doesn't look like the fascism associated with Mussolini, Hitler or Franco, 'there are many different ways to destroy democracy…[people] forget a fundamental lesson from the history of fascism: that democracy can be destroyed from within.'[25]

Fascism has never really existed in a 'pure' form – instead, there have been and remain many 'hybrid' versions: all of which need to be recognized, and recognized quickly, before it's too late to combat them. It's particularly dangerous when parties of the centre and even the centre-left make concessions to racism and bigotry. As expected, in the run-up to the 2024 election, the Tories – the most extreme version of the neoliberal 'Extreme Centre' – tried to win votes, and divide the many victims of neoliberal capitalism, via a 'toxic mix of neoliberalism and racism – banks and bigotry' aimed at stirring up fear and hatred of 'others' and minority groups.[26]

The Tory hard-right also pushed ahead with their Culture Wars against a range of human rights – yet another aspect of normalizing the ideas of 'creeping fascism'. As mentioned earlier, at least two of the Tory contenders for Johnson's job in 2022 – Suella Braverman and Kemi Badenoch – were self-identified 'culture warriors'. Their targets were refugees and asylum-seekers; rights for traveller communities; trans rights; individuals' access to human rights courts; and what they called 'wokism' in general. By October 2024, Badenoch had become one of the two remaining candidates in the Tory leadership elections to replace Sunak – with a significant lead over the other candidate, Robert Jenrik, described by *Sky News* as a 'Trump-backing voice of the right'! On 2 November, Badenoch was indeed the expected winner, giving the Tory Party its most hard-right leader ever – and showing just how far to the right Farage has dragged UK politics. Even the *New York Times* – with an extremely worrying election looming in its own country – saw Badenoch taking the Tories even further to the right. The deeply-worrying prospect of a future Tory-Reform UK coalition – or even a merger between those two parties – cannot be ruled out.[27]

Even some mainstream parties to the left of the Tories have taken up such themes in only slightly-less extreme forms – though, in essence, their stances are little different from the utterances of extreme neoliberals and bigots such as elite-members Farage and Tice, who – whilst trying to pose as 'anti-elite' populists, despite belonging to the millionaire elite themselves! – in practice are increasingly acting as outriders for full-blown fascism. Shamefully, such centre-left parties have done little to oppose 'the growth of street fascism in the fertile ground created by the mainstreaming of anti-migrant and anti-Muslim racism.' In fact, in the run-up to the 2024 election, the Labour Party often tail-ended the Tories and the populist right on migrant and refugee issues. Thus, as Faulkner had noted earlier, 'the slogans of the Far Right are picked up and waved around with the font colour changed to red.'[28]

But the far right also play a significant role in pushing back against climate policies. Often funded directly or indirectly by dirty money from the fossil fuel companies, politicians of the right have done much to push climate change denial arguments in an attempt to confuse the general public, to undermine the work of climate scientists, and to delay and weaken climate legislation. According to DeSmog, between the UK general elections of 2019 and 2024, Farage and his party received £2.3m from fossil fuel groups and think tanks funded by fossil fuel corporations. In September 2024, DeSmog exposed details of the many links between Farage and climate-crisis denying groups. Regrettably, even George Galloway and his Workers Party of Britain – which, despite its less-than-progressive stances on the climate crisis, migrants, feminism, and trans rights, nonetheless claims to be to left of the Labour Party – espoused 'red-brown' positions in face of the tidal wave of right-wing populism, xenophobia and reaction, and climate scepticism.[29]

However, Mason's call for a 'progressive' Popular Front – or 'Militant Democracy 2.0' as he calls it – to combat this growth of the right is premised on arguing, somewhat optimistically, that the 'political centre' – under the combined influences of the climate crisis and the pandemic – has now moved *away* from neoliberal capitalism! Nonetheless, he argues that a 'Popular Front remains highly relevant in the places where fascism is a threat, where the left is weak, and where liberalism is wavering.' The problem with such a Popular Front, which includes strongly pro-capitalist parties – as shown by the history of the struggle against fascism in the 1930s – is that such parties support the very system which gives rise to fascism in the first place. Hence, unlike with a more radical United Front, limits are placed on the reach of the anti-fascist struggle.[30]

One hugely important way to counter 'creeping fascism' is always to challenge its ideas which are designed to promote hate and division: whether it be xenophobia, racism, sexism, misogyny, ableism, or homophobia; and, more recently, transphobia and aggression against climate protesters or striking workers. However, one aspect of confronting – and defeating – fascism that remains important is the tactic of blocking fascist attempts to march and demonstrate. While Mason is absolutely right to argue that today's fascists spend a lot of time online, and correctly states that, because of this, 'physical resistance…works only if it's part of a wider political strategy', physical resistance nonetheless remains important. Mason himself notes that modern-day fascist groups also place a lot of emphasis on street activities and marches – including 'symbolic violence against the left, minorities and democratic institutions.' Precisely because today's fascists are skilled in their use of social media, they put a lot of effort into sharing photographs and videos of 'successful' street actions. Which is why the old tactic of 'No Pasarán!' is still so important. Essentially, fascism is a mass movement of the Right that must be confronted, because: 'It is the active mobilisation of reactionary class forces and atomised 'human dust' around the right-wing nexus of nationalism, racism, sexism, and authoritarianism.'[31]

In summary, then, 'second wave' fascism is 'an amalgam of human dust, backed by reactionary class forces, held together by myth and false promise.' The important point is that, just as the forces brought together by the Anti-Nazi League saw off the threat posed by the National Front in the late 1970s, modern-day fascism can also be defeated – and must be defeated, not just to oppose racism and intolerance and authoritarianism, but also to stop its climate-change denial attempts to undercut the attempts of the climate movement to halt capitalism's unsustainable war on the natural world. In 2019, Neil Faulkner was one of many who argued for just such a concerted effort to defeat the various aspects of 'creeping fascism': 'a united struggle from below to roll back racism and sexism, to defeat fascism, and to present a radical alternative to neoliberal capitalism and a vision of the world transformed.'[32]

Europe's 'long hot summer' of '24

Back in 2019 – before the COVID pandemic and the 'cost of living' crisis had hit – many were not convinced that the urgent call for anti-fascist unity made by Faulkner and others was a pressing concern. Nor had such doubters – despite the toxic Leave campaign of 2016 – been persuaded by the earlier warnings about the far right, issued by Michael Löwy in 2014, following the elections to the European Parliament of that year, which he saw as confirming a trend – not seen since the 1930s – that had been apparent across Europe for some years: 'the spectacular rise of the far right.' Five years later, Faulkner issued an even starker warning: 'It is necessary to recognise that we are, …engaged in a desperate, defensive, rearguard action, that we face the clear and present danger of fascism, and that we must adopt strategies of resistance and resurgence appropriate to the circumstances. *The Right is advancing. The Left is in crisis. Fascism threatens. Sound the alarm!*'[33]

Faulkner's warning of 2019 came a year after some 15,000 far-right and fascist demonstrators had attended a protest in London, called in support of Stephen Yaxley-Lennon, the fascist who is known as 'Tommy Robinson.' At the time, that demo was probably the biggest fascist gathering in British history – bigger than those of Mosley's BUF in the 1930s, and bigger than those of the National Front in the 1970s. However, even that was dwarfed by the one Yaxley-Lennon called at the end of July 2024 – in large part as a 'cultural' protest at the Trans Pride march – when an estimated 20,000 of his far-right and fascist supporters turned out to hear speeches against immigration. Then, in October 2024, another 'Robinson' call-out saw a similar far-right turn-out – and, although numbers on the anti-fascist counter-demonstration were much better than for the July demo, the anti-fascists were again outnumbered.

Those worryingly-large turnouts in the UK were just one of many alarm signals, across the whole of Europe, in the summer of 2024. The first sign that the political situation across Europe was moving further to the far right, in many countries, came with the results of the elections to the European Parliament held in June. Overall, although these elections resulted in centre-right parties remaining the biggest bloc in the parliament, far right and fascist parties nonetheless made significant gains in many countries. Particularly worrying was that a significant proportion of young voters, for the first time, swung to far-right parties. In addition, as noted by Dave Kellaway, even 'negative publicity about the Nazi sympa-thies of some leaders and plans to deport all migrants' didn't pre-vent Germany's Alternative for Germany (AfD) from *increasing* its share of the vote by three per cent. While in Italy, Meloni's Fratelli d'Italia – an essentially-fascist party – also increased its share of the vote. Perhaps such recent events in Europe – and Trump's re-elec-tion in the US in November 2024 – are early signs of an emerging collective 'shared psychosis, or *folie à millions*'? Worryingly, history does offer some examples of how, during serious crises, psycholog-ically-disturbed individuals and/or political extremists have found large audiences prepared to share their dangerous political and cul-tural views.[34]

However, probably the most worrying signs came with the elec-tions in France and the UK. In France, in the elections to the Euro-pean Parliament, Marine Le Pen's National Rally (RN) took nearly 32 per cent of the vote, winning almost twice as many votes as arch-neoliberal Macron's coalition. Macron then called a snap na-tional election. In the first round, the far-right RN came top, with over 30 per cent of the vote. To stop the far right getting into gov-ernment, the left hastily formed an electoral alliance called the New Popular Front (NFP). This was successful, in that the NFP came top, and the RN was beaten into third place. Yet, even so, the RN man-aged to win 125 seats and now, because he refuses to work with the NFP, Macron has chosen instead 'to install the RN as the arbiter of whether or not a right-wing government should be maintained.'[35]

While, in the UK – close on the back of Reform UK gaining five MPs in the general election, July 2024 had yet more far-right horrors in store. Following three fatal stabbings in Southport, Farage asked provocative 'questions' on social media – these were almost immediately followed by a far-right riot in Southport. Very quickly, far-right riots spread to other towns and cities, often with attacks on hostels where refugees were staying. In essence, those riots were just 'another example of the growing threat of fascism to draw in wider numbers of far right, populist authoritarians.' They increasingly do this by 'moderating' their language and by spreading conspiracy theories – including on the climate crisis – and rumours which are then 'amplified by fascist social media ac-counts.' By October 2024, Farage's far-right populist Reform UK had risen to 20 per cent in one opinion poll. That alone is wor-rying enough – but is especially so as Farage has made no secret of his intention to 'go after Labour', so that he can become prime minister next time round. Additionally, in an attempt to attract right-wing defectors from the Tories, Farage has also issued an ultimatum to Tory councillors, warning them to 'jump ship if they wish to avoid having Reform UK oppsosition' in the local elections due in May 2025.[36]

So: what now?

By now – if you've got this far! – you're probably feeling that Private Frazer in *Dad's Army* was right, in that not only are we well-and-truly in the shit, but that it's all so bad that in fact: 'We're doomed. Doomed!' And this is without examining the increasing frequency of nasty regional wars – such as in Palestine and Ukraine – some of which possess the potential for leading to conflict between the world's most powerful states and thus the risk of the use of nuclear weapons.

However, having established why we should indeed be experiencing 'pessimism of the intellect', the following chapters will, hopefully, gradually establish why we should also – and most importantly – embrace Gramsci's crucial 'optimism of the will.' It is that which is vitally crucial for propelling us to take the actions needed. The good news is that, essentially, it's the currently-dominant economic system, and the political elites that support it, which have created most of these crises – not humans per se. And that economic system – capitalism – has only been in existence for a very short part of human history.

While the next two chapters will show how capitalism came into existence, and why it behaves in such destructive and violent ways, subsequent chapters will show that it wasn't always like this – and will suggest how humans can get back to a more harmonious and sustainable relationship with the rest of the natural world, and with each other. Ultimately, the climate, ecological and political crises are all closely linked to the way global 'business-as-usual' is currently conducted – and it will be essential to tackle all three of those crises if we are ever to establish ecologically-sustainable and socially-just ways of living on planet Earth. Such a 'Great Transformation' seems an incredibly hard task – and it is. But it's not an *impossible* task!

BLOOD AND FIRE: BIRTH OF A MONSTER

...these newly freed men became sellers of themselves
only after they had been robbed of all their own means of
production, and all the guarantees of existence afforded by
the old feudal arrangements. And this history, the history
of their expropriation, is written in the annals of mankind in
letters of blood and fire.[1]
Karl Marx

So far, we've seen what a serious state we – and the planet – are currently in, as regards the climate and ecological crises, the threats from pandemics, and various interlinked political, social and economic crises. Most of these threats are closely associated with the way the global capitalist system operates. Yet, as with the myth that capitalism and democracy go hand-in-hand, so too are there grossly-mistaken beliefs – spread mainly by capitalists and their apologists – that capitalism came about simply because it's either 'human nature' for individuals to want to amass greater wealth than their neighbours, or because some individuals were exceptionally forward-looking, hard-working and prepared to take bold risks. The reality of capitalist development in western Europe – as opposed to such long-standing myths – is that it was mostly dependent on three main methods: enclosures and the robbery of the commons; the mass enslavement of African people; and the colonization of large parts of the Americas and Asia. As will be seen, the reality of the birth of capitalism is very far from that myth of the 'noble' capitalist entrepreneur.

Today, under the rule of capitalism, the vast majority of people have to be available – whether by choice or by force – to work for a wage, either directly or indirectly, for those who own capital. This division of society into owners of capital and workers is believed by many people to be 'normal', and even the inevitable result of 'human nature'. Yet capitalism proper has only been in existence for the last three or four hundred years or so – a *very* short time when the whole of human history is considered. More importantly, capitalism's emergence – and increasing dominance – as an economic and social system was far from 'natural'. As pointed out by Marx, capitalism only emerged as a direct result of 'a whole series of forcible methods [and] the most merciless barbarism.' And, as capitalism's more recent history has shown, such 'forcible methods and merciless barbarism' continue to be used by capitalist states and corporations to prevent attempts to establish alternative economic models.[2]

In fact, the quotation from Marx at the start of this chapter describes just one of the many violent methods associated with the birth of capitalism: the forcible robbery and privatization of the commons, along with the destruction of communal forms of agriculture across Europe, achieved by the twin processes of enclosures and engrossing. However, that forcible – and often violent – privatization of common land, and the associated denial of access to commons rights and services, was by no means the only violent method associated with the early history of capitalism. Apart from the enslavement of millions of people, and the spread of greedy and destructive forms of colonialism and imperialism, capitalism has also carried out an increasingly destructive and blind 'robbery of nature' itself. Thus, as regards the origins of capitalism, 'Human nature, if there is such a thing, had little to do with it.'[3]

Stealing the commons

Marx was quite clear on how capitalism – initially in the form of agricultural capitalism – came into being: 'Expropriation is the starting-point of the capitalist mode of production, whose goal is … to expropriate all individuals from the means of production.' To underline the violence of the birth of capitalism, Marx also noted how capital came into the world 'dripping from head to toe, from every pore, with blood and dirt.'[4]

Marx was only one of many who established that one of the key factors in the birth of European capitalism was the process of enclosure: the privatization of land worked in common, and of the rights and services associated with the existence of common lands. The latter included rights – for even the poorest people – to gather firewood, pick nuts and berries, fish and hunt, and to glean any grain leftover after harvesting had been completed. All these means of subsistence allowed a certain amount of independence – and hence meant there were a large number of people for whom working for someone else, for a wage, was not vital for survival. A poetic account of the impact of enclosures on rural communities is provided by the English poet, John Clare (1793-1864):

> *Inclosure came and trampled on the grave*
> *Of labour's rights and left the poor a slave...*
> *And birds and trees and flowers without name*
> *All sighed when lawless law's enclosure came.*[5]

To provide a large wage-labour force for those wealthy land-owners who were keen to extract capital from their farms, common land and those common rights had to be destroyed – with all access to such services then becoming the exclusive preserve of large property owners. In England, this process began as early as the end of the fifteenth century as the feudal system increasingly began to crumble, and was achieved via a combination of political dominance, fraud, and naked violence. By splitting what had traditionally been shared fields into private plots, separated from each other via expensive fences or hedges, and by enclosing the commons, an ever-increasing number of formerly-independent commoners were pushed by the need to survive into working for those private land-owners who had 'appropriated' common land and rights.

As Marx noted about the rise of private property and what has been called the 'primitive accumulation' of early capital: 'In actual history, it is a notorious fact that conquest, enslavement, robbery, murder, in short, force, play the greater part.' Some – such as Ian Angus – argue that Marx's use of the term 'primitive accumulation' here is best replaced by the phrase Marx preferred: 'original expropriation.' The latter certainly gives a more complete idea of the theft and violence involved. The force used soon resulted in most rural people being increasingly separated from their earlier means of subsistence, with 'great masses of men' being suddenly and forcibly torn from their means of subsistence, and hurled onto the labour-market as free, unprotected and rightless proletarians. The expropriation of the agricultural producer, of the peasant, from the soil is the basis of the whole process.'[6]

During the fifteenth century, some landlords began to concentrate on producing wool for the Flemish cloth industry. Consequently, they forcibly evicted smaller tenants, consolidated those lands, and then leased these larger farms – at increased rents – to wealthier farmers, and increasingly to individuals and commercial companies wanting to invest in sheep. The attraction of sheep farming – apart from the high prices paid for wool – was that it needed less labourers than growing grains and other crops. By the early sixteenth century, this process – as well as forcing many smaller tenants into the ranks of landless wage labourers – had also led to rural depopulation and even to the disappearance of many villages. It was that which led Thomas More, in 1516, to write in his book *Utopia* that sheep had 'become so greedy and fierce that they devour human beings themselves. They devastate and depopulate fields, houses and towns.' Though of course the depopulating effects of enclosures weren't down to 'greedy sheep' – they were down to the increasingly-greedy, large capitalist landowners. Often forcibly evicted at short notice, former tenants were pushed into becoming wandering 'vagabonds' looking for often non-existent work, sometimes resorting to begging or theft in order to survive – both of which were illegal and carried harsh penalties.[7]

Although various laws were passed during the reigns of both Henry VII and Henry VIII, to slow – and even halt, – the enclosure processes, landlords either simply ignored them completely, or found ways of circumventing them. In fact, Henry VIII actually made the situation worse via his dissolution of the monasteries and the subsequent selling of their massive landholdings. These lands were often bought by various types of businessmen determined to maximise profits from their new farms: 'The Tudors didn't just fail to halt the advance of capitalist agriculture, they unintentionally gave it a major boost.' By the mid-sixteenth century, the number of 'able-bodied poor' had increased significantly. The resulting anger felt by rural populations as the rate of enclosures increased, and agricultural capitalism spread – forcing many off the land and into the ranks of wage-labourers – eventually exploded into violent resistance.[8]

Popular resistance

Probably the most famous example of such resistance to enclosures and the robbery of the commons took place in Norfolk in 1549. This was Kett's Rebellion: a widespread social uprising and revolt against those processes. Significantly, the first of the rebels' *Twenty-nine Demands* dealt with enclosures: 'We pray... that from hensforth no man shall enclose any more.' While the third demand stated: 'We pray...that no lord of no mannor shall com on upon the comons.'[9]

The Norfolk rebels set out for Norwich – at that time, the second-largest city in England – from the small market town of Wymondham. As they went, they tore down hedges and fences that had divided formerly common land into privately-owned fields and pastures. Eventually, some 20,000 rebels camped on Mousehold Heath outside Norwich – and, on two occasions, the rebels captured Norwich itself. However, they were finally defeated by a professional mercenary force numbering some 4,000. Following their defeat, over 3,500 rebels were massacred, and their leaders tortured and then beheaded.

In fact, the year 1549 turned into one of the most turbulent in the entire history of Tudor England – and was remembered for many years as the 'Commotion Time'. By the late summer, nearly half of England was in open revolt against the government. There were similar petitions, protests and rebellions in at least 25 counties which, according to Fletcher and McCulloch, showed 'unmistakeable signs of coordination and planning right across lowland England.' Initially, given the number of uprisings, and the number of rebels involved, the local gentry and the authorities 'were powerless in the face of [those] massive demonstrations.' Ian Angus sees those rebellions of 1549 as demonstrating 'as nothing else can, the devastating impact of capitalism on the lives of the people who worked the land in early modern England' – and as verifying what Marx called 'the systematic theft of communal property [in which] the place of the independent yeoman was taken by tenants at will, small farmers on yearly leases, a servile rabble dependent on the arbitrary will of the landlords.'[10]

Eventually, the traditional rural economy was destroyed by those two processes, with wealthy landlords grabbing the commons for themselves, and some farmers looking to increase yields and produce surpluses that could be sold to wider markets for profit. This was achieved by small-holders either being evicted from the land they rented, or being forced out of common-field villages by other means. Although term 'enclosures' was usually used to describe this change, 'the process was more complicated than just erecting fences.' The change from a communal system of farming in open fields to one based on individual farming resulted in 'the extinction of common rights.' … The other threat to the independence of small tenants was 'engrossing', which involved 'the absorption of entire farms into larger units, usually for sheep farming.[11]

Nonetheless, resistance to those capitalist developments – and rejection of the ideology underpinning them – continued. During the English Revolution of the 1640s, a truly radical group, known as the Diggers – though they called themselves the 'True Levellers' – took radical non-violent action. Their main spokesperson was Gerrard Winstanley, from Wigan in Lancashire. In April 1649, they began digging uncultivated land at the side of St. George's Hill in Surrey, with the intention of growing corn and other foodstuffs. They told the local inhabitants that they intended 'to open and present the state of community to the sons of men' and 'to prove that it was an indeniable equity that the common people ought to dig, plow, plant, and dwell upon the Commons without hiring them, or paying rent to any.' Initially, they were wrongly accused of favouring the 'doctrine of Parity or levelling, bringing all mens estates to an Equallity.' They, however, made it clear that 'they intended not to meddle with any man's property, nor break down pales or fences…but only to meddle with what is common and untilled.'[12]

Later, in his revolutionary pamphlet 'The New Law of Righteousness', Winstanley challenged the prevailing view that: 'the Earth was made for a few, and not for all men.' Winstanley and the Diggers believed that 'property' meant liberty for the few who possessed, and slavery for the masses. What they wanted was for the Earth, once again, to become a 'Common Treasury', so that everyone would have enough to eat – while poverty, working for a wage, and 'buying and selling', would disappear. If the wealthy objected to sharing their lands, the Diggers' view was that they could work their land with their own hands, while the 'common people' worked the common lands together. Though their views were in large part based on their religious beliefs, the various pamphlets that Winstanley produced – especially his 'The Law of Freedom' – came very close to the communistic thinking of Marx and of nineteenth-century French socialists. In particular, Winstanley argued that as no one could become wealthy without the 'help' (i.e., work) of others, then the riches gained by wealthy individuals should be shared with those who'd helped produce that wealth. Thus, according to one early historian of Winstanley and the Diggers, their ideas promised 'to ultimately produce a change in social conditions compared with which the abolition of slavery sinks into comparative insignificance.'[13]

Not surprisingly, the Diggers' attempts at working the commons were quickly crushed by the authorities and, during the second half of the seventeenth century, laws allowing enclosures began to be passed. By then, those sitting in parliament increasingly came from the ranks of more 'enterprising' – i.e., capitalist – landowners. As a consequence, the rate of enclosure accelerated, and the poverty experienced by smaller farmers and cottagers increased. Another development – which also made it more and more difficult for independent smallholders and the rural poor to survive – was wealthy people enclosing woods and land for their exclusive right to hunt via a succession of Game Laws which, like the infamous Black Act of 1723, made 'poaching' a serious criminal – and in certain cases, a capital – offence.

At the same time, a capitalist ethos – which pitted man against man, as opposed to the paternalism associated with the old feudal order – began increasingly to take root. This new, capitalist, ideology was based on the concept of what Macpherson called 'possessive individualism.' By this, 'The individual was seen neither as a moral whole, nor as part of a larger social whole' – instead, 'society' was defined in two ways: firstly as 'a lot of free, equal individuals related to each other as proprietors of their own capacities and of what they have acquired by their exercise'; and secondly merely as 'relations of exchange between proprietors.'[14]

This essentially capitalist worldview was premised on the belief that those who own property have an absolute right to do as they wish with whatever property they've been able to 'acquire.' This attitude towards privatization of the commons has, over the succeeding centuries, gone beyond being centred on land, to become the prevailing attitude of neoliberal capitalism regarding the privatization of *all* public services. At the time, however, there was continued strong resistance to being forced into the ranks of wage-labourers – who were often described as 'servants' and thus seen as 'unfree'. As Angus explains: 'Not only were wages low and working conditions abysmal, but the very idea of being subject to a boss and working under wage-discipline was universally detested.'[15]

For those forced off the land in the more prosperous rural areas, one way of escaping the dreaded necessity of becoming a wage-labourer – or, as Marx described them, 'free, unprotected and rightless proletarians' who had been 'suddenly and forcibly torn from their means of subsistence, and hurled onto the labour-market' – was to move to the rural margins, such as heathlands, forests, marshes and fens. There were still extensive commons in such areas, and people could squat there in huts and establish small farms. This, along with common rights and occasional wage-work, allowed families to survive whilst still retaining some independence. But, as such areas were increasingly enclosed and privatized, many were eventually forced into migrating – to the Americas, to coalmining areas, or to larger centres of population where wages were higher.[16]

The transition to capitalist agriculture was thus both a brutal, and a long drawn-out, process. Nonetheless, by the early eighteenth century, about 30 per cent of England (and almost all of Scotland) was still unenclosed, and most people still lived and worked on the land. Consequently, the capitalist 'war against the commons' was stepped up. As Angus has noted: 'It took another great wave of assaults on commons and commoners, after 1750, to complete the transition to industrial capitalism…Over time, and with many detours and reverses, the dispossessed became proletarians.'[17]

The triumph of 'possessive individualism'

By the eighteenth century, the larger farms, created via the enclosure of common lands, were increasingly being called 'capital farms, or merchant farms.' At the same time, the process of consolidating farmland had significantly increased the number of landless workers who, according to capitalist ideology, were now 'free' because they were no longer tied to a particular landlord, and instead were 'free' to sell their labour power to any employer. Of course, however, they were also 'free' to starve if they couldn't find an employer, or gain a living wage. By '"setting free" the agricultural population as a proletariat for the needs of industry', such developments were important steps towards the emergence of industrial capitalism.[18]

As Marx and Engels noted in *The Communist Manifesto*, the result of all these mostly-forced changes was that 'private property is already done away with for nine-tenths of the population; its existence for the few is solely due to its non-existence in the hands of those nine-tenths.' As a result, it has been calculated that, by the early nineteenth century, the richest 1 per cent of the population owned 49 per cent of the land, while the richest 5 per cent owned 86 per cent. But more was needed than just driving the less well-off from the land to get them to work for wages for others. Essentially, the process of enclosures was accompanied by a deliberate policy of using hunger to force people to work for others.[19]

One way, alongside laws against begging and laws to force the poor to work, was to restrict – and even abolish – all forms of public relief for the poor, and thus let hunger drive the dispossessed poor into working for wages. As the nineteenth century 'progressed', while emerging capital class lauded the principles of free choice and free markets, they had a very different attitude when it came to the poor. On the contrary, they believed 'that the rural poor should be forced to work for them' and not be allowed to work for themselves – because 'the more they work for themselves, the less they will for us.' This was a brutal but clear expression of the fact that the success of capitalist farming was dependent on ensuring a sufficient and regular supply of farm labourers. They specifically rejected allowing poor people to settle on what commons still remained, and thus benefit from commons rights, in case 'it operates upon their minds as a sort of independence.' The worry that commoners would become too independent to work for large landowners was voiced by a member of the Royal Society: 'If you offer them work, they will tell you that they must go to lock up their sheep, cut furzes, get their cow out of the pound, or perhaps they must take their horse to be shod, that he may carry them to a horse-race or cricket-match.' While another warned that, if a common man were allowed access to more land than he and his family could work in the evenings, or at other 'leisure' times, then 'the farmer can no longer depend on him for constant work, and the hay-making and harvest (without very favourable weather) must suffer to a degree which (in extent) would sometimes prove a national inconvenience.'[20]

The capitalist farmers – who clearly thought any 'leisure' time for poor people should be spent working on their tiny plots – did not apply the same standards to themselves. As Angus points out, their 'blatant and crude' class bias was both made absolutely clear, as they never 'suggested that owning thousands of acres made the gentry and aristocracy indolent and let them live without working.' Though, of course, that is precisely what it did to many of the one per cent. In fact, many landowners didn't even want the rural poor to have their own gardens, let alone be able to have access to what remained of the commons, because this also led to the: 'disinclination on the part of the cottier [cottager] to be employed.' The same attitude emerged amongst those with capital to invest in the newly-emerging industries. Many expressed concerns that factory work was considered so uncongenial to most urban workers that it wasn't always possible to keep the machines working for twelve or fourteen hours a day. In the early days of the Industrial Revolution, most factory workers were women or children – or inmates of workhouses who were forced to work as little more than slaves. This is why the early factories looked like prisons: they had high barred windows, and doors were locked: because early industrial workers often felt they'd earned enough to address their needs *before* the end of the working week, and thus left work early, not returning until the following week. Additionally, penalties – for lateness, and even talking – were imposed in the factories to coerce workers into working longer than they would otherwise have chosen to do. Something today's firms – such as Amazon – have clearly adopted from the nineteenth century! Thus: 'Under the spur of hunger, men and women who had largely controlled the pace and intensity of their work, who had worked collectively and convivially with their families and neighbors, were made subject to the dictates of capital.'[21]

The robbery of nature

As well as robbing the common lands from the majority of the population, capitalism has also robbed nature itself – from its earliest phases. This was done by making land 'subject to the dictates of capital' – initially, by robbing the soil of nutrients. Marx was aware of this early on; in fact, as noted by Foster and Clark: 'Marx's notion of the robbery of the soil is intrinsically connected to the rift in the metabolism between human beings and the earth' that he saw being created by capitalism. Marx had been alerted to the increasingly destructive nature of capitalist agriculture by the work of Justus von Leibig, a German chemist. This led him to write that all 'progress' in capitalist agriculture was not just based on robbing the worker, but also on robbing the soil; and to conclude that 'all progress in increasing the fertility of the soil for a given [short] time is a progress towards ruining the more long-lasting sources of that fertility.' Such robbery resulted from capitalism disturbing 'the metabolic interaction between man and the earth' in such a way that, by preventing the return to the soil of the 'constituent elements consumed by man in the form of food' it thus hindered 'the operation of the eternal natural condition for the lasting fertility of the soil.'[22]

Marx returned to this theme later on, when commenting on how capitalism – first in its agricultural operations, and then in its industrial phase – by driving people from the land into factory towns, produced conditions that resulted in 'an irreparable rift in the interdependent process of social metabolism, a metabolism prescribed by the natural laws of life itself.' As a result, there occurred 'a squandering of the vitality of the soil, which is carried by trade far beyond the bounds of a single country.' Thus, although the industrialization of agriculture initially led to a period of increased yields – especially after the repeal of the Corn Laws – as farmers were constantly striving to increase yields in order to survive market forces, it already contained 'deep ecological and economic contradictions, threatening the future of British agriculture.' Essentially, 'a metabolic rift, caused by the intensive robbing of soil nutrients and a boom-and-bust cycle, was built into industrial-capitalist agriculture.'[23]

In effect, capitalist agriculture destroyed the 'vitality of the soil' because it 'was not motivated by sustainability but by increasing production and capital accumulation.' A 'high farming' system developed that was increasingly marked by the need for ever-more applications of fertilizers. The result was a capitalist agriculture that was massively dependent on 'energy-intensive inputs.' In fact, a soil fertility crisis developed as early as the 1830s and 1840s, because of 'the lack of fertilizers to replace the nutrients shipped as food and fiber to the cities and thus lost to the soil.' Far from the so-called 'Agricultural Revolution' making the UK self-sufficient in foodstuffs, as was originally argued by economists and historians, yields began to decline. After 1740 – following almost two centuries of fairly consistent availability of food, food production per person 'fell dramatically.' From the mid-eighteenth century to the early nineteenth century, there was only a ten per cent increase in agricultural production: 'essentially zero annual growth.'[24]

The results were two-fold. Firstly, during English agriculture's so-called 'golden age', most agricultural and industrial workers barely consumed enough to avoid diseases associated with starvation, or even to survive. This was particularly true at times of bad harvests and thus rising food prices. Similar to the anti-enclosure uprisings in the sixteenth century, the late eighteenth century saw frequent and widespread food riots which demanded the right to buy food at affordable prices. Something, of course, which ran directly counter to the capitalist 'free market' ideology which had come to dominate the economy. Consequently, workers in the late eighteenth and early nineteenth centuries were 'malnourished: they were less healthy and died younger than their ancestors a century earlier.' This has been confirmed by a global study by Dylan Sullivan and Jason Hickel, published in 2023, which concluded that 'the rise of capitalism from the long 16th century onward is associated with a … deterioration in human stature, and an upturn in premature mortality.' Secondly, from about 1750, Britain moved from a country that was self-sufficient in food and able to export food, to a country that increasingly needed to import food, especially grains. This was in large part because capitalist farmers saw more profit in shifting from grain to meat and wool production, as the latter required far fewer labourers. It became increasingly clear that the main object of the later enclosures and theft of common lands was not, as was often claimed to increase food production, but to maximise profits. Thus, as ever with capitalism: 'Profits took priority over people.' This food-gap was filled by increasing imports from Britain's colonies, and from countries such as Germany, Russia and the US. This in effect meant that 'a large part of the British metabolic rift was transferred abroad' – in what has been described as 'a radical shift to outsourcing.' By the early twentieth century, it's been estimated that over 80 per cent of the bread consumed in the UK came from imported grain.[25]

Thus, in effect, nineteenth-century agricultural capitalism was soon producing similar effects far beyond the bounds of Britain itself. This can first of all be seen in the emerging robbery of fertilizers – and ultimately soil fertility – from various countries, mostly but not exclusively, in the global South. This was because capitalist agriculture was increasingly destructive, owing to it being based, above all, on creating value and accumulating yet more capital. This meant that, despite increasing capital investments, capitalism's 'rapacious expropriation of nature inevitably promoted ecological destruction.' This was why Marx had stressed that capitalism was 'the opposite of an ecologically self-sufficient system.' Because capitalist agriculture in England had focussed on vastly increasing productivity, and had thus significantly destroyed the soil metabolism, it was soon forced to seek fertilizers from further afield. This operated even within the British Isles, in that Irish agriculture – largely in the hands on absentee English landlords – saw Irish manure in effect 'exported' to England via the export of grain, in what Foster and Clark have described as 'an early form of unequal ecological exchange.'[26]

This was particularly brutal when it came to the extraction and export of guano (bird dung) on the islands off the coast of Peru. To obtain this, large numbers of bonded workers – known as 'coolies' – were shipped in from China to dig this natural fertilizer. Between 1847 and 1850, England imported 88,540 tons of guano – by 1868-71, this had risen to 209,460 tons. However, this increase far exceeded 'the growth rate of agriculture.' In addition, the conditions of work on those islands were so awful that 'the suicide rate of the Chinese bonded workers digging the guano was so high' that, as the US consul to Peru reported in 1870, guards had to be employed to stop those workers throwing themselves into the ocean in despair. English agriculture also came increasingly to rely on the import of bones dug up from the battlefields of Europe and beyond.[27]

Essentially, capitalist agriculture was robbing nature so much that it 'required massive energy inputs in the form of fertilizers and material inputs from abroad that increased far more rapidly than did productivity in agriculture. It also required, even in the heyday of high farming, the massive and rapidly increasing import of wheat.' This was because capitalist agriculture, by separating the bulk of the population – and their sewage – from the land, caused major disruptions (or, to use Marx's phrase, 'metabolic rifts') in the soil nutrient cycle. Typically, instead of trying to make agriculture more ecologically sustainable, the rifts were widened even further by the increasing use of fertilizers and, later, herbicides and pesticides. Whilst offering some short-term solutions, the long-term result was to increasingly transgress ecological limits. Or, as Marx put it, capitalist agricultural production inevitably undermined 'the original sources of all wealth – the soil and the worker.' However, it's hugely important to note – especially in view of the rationale behind the ecosocialist project – that Marx believed that, despite the 'irreparable rift' created by capitalism, 'a restoration of a rational and sustainable metabolism between human beings and the earth' was perfectly possible: provided capitalism was replaced by what was then called 'socialism', but which now increasing numbers of people understand to be *eco*socialism.[28]

Capitalism and enslavement

Another – and earlier – extremely violent aspect of the birth of European capitalism was the enslavement of literally millions of people, mainly affecting peoples living on the African and American continents. While slavery – sometimes widespread – was a feature of earlier empires (such as the Roman Empire), the form and scale of what became known as the transatlantic 'slave trade' had very clear capitalist features from the start.

As David Whyte has shown, investors in early eighteenth-century Britain – such as those who got involved with short-lived South Sea Company, formed in 1711 to expand British control of the trafficking of enslaved people from the Caribbean and North America to Spanish-controlled Latin America – 'were rubbing their hands at the prospect of a major expansion in the human slave trade.' Investors knew full well that they would be profiting from a scheme based on the exploitation of human victims. In the case of the South Sea Company, for example, the 'investors' – who wouldn't have liked to be described as 'slave traders' – were hoping to ship 145,000 enslaved Africans to Latin America. This kind of 'cognitive dissonance' is particularly associated with those who run capitalist corporations, and those who invest in them. According to Whyte – among others – 'the corporation enables capitalists to disregard moral limits or indeed any limits on their profiteering that we might expect to be consistent with a basic sense of humanity.'[29]

In fact, several trading companies were formed to oversee and develop the African slavery 'market'. For instance, the Royal African Company was formed in 1662 by the British Crown and merchants from the City of London. Though originally planned to trade and extract gold on the west coast of Africa, it was later granted a monopoly to trade enslaved people. Thus, within twenty years of its formation, it was also 'trading' some 5,000 enslaved people every year. Although it eventually lost its 'slave monopoly' in the early eighteenth century – because it was then decreed that trading in enslaved people was 'a fundamental and natural right of [all] Englishmen' – by then, many traffickers in enslaved people had learned the 'skills' needed to succeed in that 'trade.'[30]

In all, as Mikaela Loach has pointed out, capitalism's brutal and barbaric 'slave trade' captured, stole and enslaved 'an estimated thirteen million people...for profit.' The wealth generated by this forcible and brutal enslavement of millions was quickly invested in the imperialist homelands, initially in agricultural land and then, increasingly, in the early stages of the Industrial Revolution. But even after the obscenity of the transatlantic trafficking in enslaved Africans – and then, later, 'slavery' itself – was abolished, robbery and violence continued to be inflicted on the peoples of Africa, and elsewhere, via colonialism and imperialism. Later, this was then replaced by capitalist neo-colonialism and neoliberalism. [31]

The murder connection: capitalism and colonialism

A traditional comment about the British Empire – which encompassed about 25 per cent of the world's land surface at its peak – was that it was an empire 'on which the sun never sets.' In 1851, Ernest Jones, the radical Chartist, gave a more complete view of the reality of the British Empire, by adding the words: 'and on which the blood never dries.' That blood of course was part of the price paid by the countless victims of British colonialism's global theft, fraud, violence and murder. It was also the price paid by the mainly working-class troops sent to steal and then secure those lands for various capitalist corporations which enriched themselves at the expense of everybody else. What was true of British colonialism was true of European colonialism in general – including Spanish, Portuguese, French, Italian, Dutch, German and Belgian imperialism. So severe was the historical exploitation inflicted by those capitalist states and corporations that even today, 'we often refer to many previously colonised nations as "poor" [though] without giving full context as to why that is.' Yet the exploitation of ex-colonial countries by capitalist corporations continues even today. However, such countries are not 'poor' because they are 'under-developed', but poor because they were, and continue to be, *over-exploited*. Additionally, such historic and contemporary exploitation has left the peoples of these countries – even though they have done least in terms of CO_2 emissions – 'more vulnerable to climate breakdown.' This is because the 'extraction of wealth from these nations and the deliberate underdevelopment of them' means they now lack 'the capacity to create the necessary infrastructure to survive the impacts of the climate crisis.' Hence the on-going calls for restitution and compensation being increasingly raised by the climate justice movement.[32]

As has been seen, capitalism as we know it really began from the sixteenth century onwards – and, in Britain and other industrializing European nations, was associated with a decline in living standards for the majority, to the point of falling below subsistence levels. But such impacts were even more pronounced in those countries in Africa, Asia and Latin America that increasingly came under the yoke of

European colonialism. So much so that, even today, many of those countries are still experiencing low living standards, poorer health, and shorter lives. In a recent paper on the global impact of capitalism, *Capitalism and extreme poverty*, Sullivan and Hickel have convincingly shown that 'the rise of capitalism coincided with a deterioration in human welfare.' This was true not just of Europe, but also for those global South countries colonized by European nations: 'In every region…, incorporation into the capitalist world-system was associated with a decline in wages to below subsistence.'[33]

From the late sixteenth century onwards, it was capitalist corporations – such as the East India Company – which were the main operators in the colonial expansion of European powers. It was chiefly such corporations which carried out the theft and enclosure of traditional lands; the organization firstly, of slavery and then later of the slave trade; and the economic exploitation and commodity-extraction of the newly-colonized people and lands. To enforce its wealth-gathering operations, such corporations were often granted the right, by their home governments, to raise their own armed forces – and thus the legal right to use violence against any who resisted. Or, as Rosa Luxemburg put it in 1913, capitalism must, by its very nature, 'always and everywhere fight a battle of annihilation against every historical form of natural [non-capitalist] economy that it encounters.' In the countries of the global South, that 'battle of annihilation' 'assumes the forms of colonial policy.'[34]

Early tales – used to justify empire-building – that extreme poverty had been the 'natural state' of the newly-colonized countries, and that it was the advent of capitalism which reduced such extreme poverty, are false. Instead, the reality was that 'extreme poverty was uncommon and arose primarily during periods of severe social and economic dislocation, particularly under colonialism.' Furthermore – and of real significance as regards the history and impacts of capitalist colonialism and imperialism in the global South – Sullivan and Hickel have shown that the significant improvements in human welfare which began in such countries during the second half of the twentieth century were mainly achieved by *defying* capitalism: 'These gains coincide with the rise of anti-colonial and socialist political movements.'[35]

One of the main reasons why colonialism created such poverty was linked to how capitalist agriculture in Britain quickly robbed the soil of its nutrients, thus making the UK increasingly dependent on imported fertilizers and grains. As early as 1848, the philosopher and economist John Stuart Mill had commented on how England was no longer able to depend on the fertility of its own soils, and, instead, had become dependent on 'the soil of the whole world.' Or, to put it another way: 'England avoided mass starvation during the Industrial Revolution by robbing the soil and starving the people in its imperial possessions.' The concept of 'slow violence' was recently used by Chris Otter (in his 2020 book, *Diet for a Large Planet*) to describe what Britain frequently inflicted on its colonies – such as Ireland and India – by continuing to export food to the imperial homeland even during periods in which local harvests failed and famines began to develop. Some estimates conservatively put the combined death toll during the Irish famine of 1845-50, and the three main Indian famines between 1876 and 1945, at around 13 million people. As Otter summarized it: 'Famine was effectively outsourced.'[36]

In Ireland, with over 50 per cent of land owned by absentee English landowners by the mid-eighteenth century, the bulk of the agricultural products produced in Ireland were shipped to England – leaving millions of Irish cottagers and labourers to survive on potatoes and skimmed milk. By 1845, 60 per cent of Ireland's agricultural produce was being exported – while profits also went to England, instead of being used to replenish the exhausted soil with nutrients.

Other colonial or semi-colonial areas, such as the West Indies and China, were increasingly given over to the production of sugar and tea, respectively. While by the end of the nineteenth century, most of the wheat grown on over 25 million acres in India was exported to Britain. This was because, as argued earlier, the myth of capitalism's so-called Agricultural Revolution had failed to feed Britain's labouring classes. In effect, the amount of food that was produced in the UK proved inadequate to allow for the feeding of workers driven from rural areas into the factory towns of capitalism's early Industrial Revolution. Only increasing amounts of imported food and raw materials, sucked vampire-like from British colonies, made Britain's nineteenth-century transition to a largely industrial society – and, later, to a capitalist state with a welfare system – possible. Added to all of this is the issue of racism, which was *always* an essential element in, and result of, colonialism – as it was in the slave trade. Racism has also often been an important element in capitalist corporations pursuing the 'logic' of ever-more capital accumulation. As Whyte put it: 'Corporate racial exploitation is buried deep in the foundations of the history that made capitalism.'[37]

The impact on those countries colonized was unsurprisingly dire – entire communities were often driven from the land and displaced to allow for mining or the production of 'cash' crops; while those 'who had lived off the land were now forced into exploitative wage labour in order to survive as the land was destroyed and the environment poisoned by pollution.' It's estimated that at least $45 trillion was extracted from 'British' India alone, to be used in financing industrialization in Britain. This was achieved by food and raw materials being shipped from India to Britain, which then exported those goods across the world, with the profits remaining in Britain. As Loach summarizes: 'The industrialisation and so-called development of Britain relied upon the de-industrialisation of India.' For many, though, European colonialism – including British imperial expansion – is often associated with what became known as the 'Scramble for Africa' which took place during the last two decades of the nineteenth century. The race to grab colonies in Africa – and, at the same time, large parts of Asia – was also an important factor in the outbreak of the First World War. In 1913 – at the culmination of this process – Rosa Luxemburg traced the connections between capitalism's robbery of the commons in Europe to the global development colonialism and imperialism: 'dispossessing the peasants in England and on the Continent was the most striking weapon in the large-scale transformation of means of production and labour power into capital…capital in power performs the same task even to-day, and on an even more important scale – by modern colonial policy.'[38]

Of course, capitalist corporations – via the global spread of their neoliberal tentacles – still continue to profit, massively, from such under-developed countries via more modern forms of robbery and exploitation. In fact, it has been argued by both Jeremy Walker and Jason Moore that 'the rise of the economic thinking, structures and fortunes accompanying the formation of neoliberalism all have their roots in colonial domination and exploitation.' The important point to grasp in all this is that, in many cases, poverty is *created* by capitalist corporations as a direct result of their global operations. Quite simply, poverty isn't a problem that just 'exists.' As Jason Hickel established in his book, *The Divide* (2018), the income gap between the global North and the global South has roughly *tripled* in size since 1960. Today, as a result, 60 per cent of the world's population – around 4.3 billion people – live on less than $5 per day whilst, at the same time, 'the richest eight people now control the same amount of wealth as the poorest half of the world combined.'[39]

Essentially, the poorer countries in the global South are poor for two main reasons: firstly, they have been plundered and their development held back by their experiences of colonialism; but secondly, they remain relatively poor because they have been, willingly and unwillingly, integrated into the global capitalist system on markedly unequal terms. Even so-called 'aid' masks continued and massive wealth extraction which continues to cause real poverty. Large predatory corporations are still grabbing land – sometimes violently – from local farmers and communities; while trade deals are usually massively-rigged in favour of the big transnational corporations, which also often manage to avoid paying taxes where they operate. In addition, as recent years have shown, such countries are now having to deal with the increasingly-destructive human and ecological costs of capitalist-induced climate change. As Kate Raworth has said of Hickel's book: 'There's no understanding global inequality without understanding its history.' And that history is why colonialism continues to be a 'centuries-long crime' – and a crime that needs to be put right.[40]

By now, it should be abundantly clear that capitalism is not a 'natural' economic system – in any sense of the term. Instead of being 'natural', capitalism is instead a specific social system that has arisen as the result of the greed of a minority – *not* by the conscious choice of the majority. The next chapter – sadly, yet another 'doom and gloom' chapter – will examine how and why capitalism has now become dangerously out of control.

OUT OF CONTROL

[The] design [of the GDP growth] model is fundamentally flawed because it runs counter to the living world, which thrives by continually recycling life's building blocks such as carbon, oxygen, water, nitrogen and phosphorus ... [the GDP model breaks] these natural cycles apart, depleting nature's sources and dumping too much waste in her sinks.[1]
Kate Raworth

So far, the six preceding chapters have been concerned with establishing the increasing extent, and mounting seriousness, of the multiple crises we currently face – and with pinning the bulk of the blame for all those crises on the main culprit: capitalism. This chapter will focus on how neoliberal capitalism itself is now dangerously 'out of control' – with both capitalist corporations and states increasingly unable, or unwilling, even to limit some of its worst excesses, never mind stop them and begin restoring the natural world and social justice.

Given that we've now seen just how brutal and destructive the birth of capitalism was, and how this new economic system trampled (often quite literally) over the majority of the population in order to become established, it shouldn't be surprising that capitalism – despite all the warning evidence of what its operations are now doing to people and planet – is now well and truly set on a destructive course it's unable to stop. It is thus no coincidence that, over the decades, terms such as monster, vampire, juggernaut and vulture have been used by critics of capitalism to describe its impacts on people and planet.

One of the main reasons for capitalism's determination constantly to override scientific evidence and, instead, persist in its deadly ecocidal operations, is what has been called its 'destructive uncontrollability.' This is closely linked to its constant drive for ever more GDP 'growth', via expanding extraction, production and consumption, in order to increase profit and capital accumulation for the capitalist class. For, in the end, it is this very small minority of the world's population who, because they 'own' most of the world's resources, get to make the important decisions.

'Destructive uncontrollability'

Some people are surprised to discover that Marx and Engels once acknowledged, to some extent, how creative capitalism had been since its first appearance. As early as 1847, in *The Communist Manifesto,* they wrote that in its first hundred years or so, capitalism had 'created more massive and more colossal productive forces than have all preceding generations together... — what earlier century had even a presentiment that such forces slumbered in the lap of social labour?'

When they wrote this, only a small part of the world had come under the sway of capitalism. Since then, of course, capitalism has become a truly global economic system, dominating virtually every country in the world. However, alongside its undoubted creativity, capitalism has also – and always – been marked by crises and destruction. This increasingly-fatal duality of capitalism – especially in relation to the natural world – was noted by the Marxist economist, Paul Sweezy, in 1989: 'Implicit in the very concept of this [capitalist] system are interlocked and enormously powerful drives to both creation and destruction.' The most serious aspect of capitalism's destructiveness is that it 'bears most heavily on nature's capacity to respond to the demands placed on it.'[2]

Even supporters of capitalism – such as the Austrian economist, Joseph Schumpeter – have recognized that capitalism is a destructive as well as a creative economic system. However, instead of seeing its operations as 'creative destruction', critics of capitalism – such as John Bellamy Foster – have argued instead that '*destructive* creativity' is a much more apt description: 'Capital's endless pursuit of new outlets for class-based accumulation requires for its continuation the destruction of both pre-existing natural conditions and previous social relations.' This is why the exploitation of people, imperialism, war, and ecological devastation are interrelated and 'intrinsic features of capitalist development' – thus it's important to grasp that such aspects are 'not mere unrelated accidents of history.' Others – such as István Mészáros, in his 2001 book *Socialism or Barbarism* – have pointed to the danger that capitalism's 'destructive creativity' would eventually turn into what he described as 'destructive uncontrollability.' Essentially, he argued that the destruction – which is at the heart of capitalism's 'logic' of continually-expanding production and capital accumulation – would eventually escape control of both capitalist corporations *and* governments, so that the conditions of life itself, for humans and all other species, would be increasingly destroyed, thus threatening human civilization itself. For an increasing number of people today, it has now become worryingly-clear that 'such destructive uncontrollability has come to characterize the entire capitalist world economy, encompassing the planet as a whole.'[3]

As well as having been compared, by Marx and others, to a 'vampire' draining life from the natural world, and from the humans it exploits, capitalism has also been seen as an out-of-control juggernaut. This term was used by Sweezy, who argued that since its birth, capitalism has been 'a juggernaut driven by…small groups single-mindedly pursuing their own interests.' While in the short-term, capitalism is to some extent controlled by market forces, those market forces tend to fail in the longer run, leading to 'devastating crises.' This is because of capitalism's central 'logic', which pushes it ever-forwards in a never-ceasing drive for economic expansion: not in the general interests of humanity as a whole, but simply for ever-greater profits and the accumulation of massive amounts of wealth for the small minority who ultimately own this economic system. As noted by Sweezy: 'It is this obsession with capital accumulation that distinguishes capitalism from the simple [or natural economic] system for satisfying human needs.' As was seen in the previous chapter, it was clear from early on that capitalism was a system that exploited (often ruthlessly) humans for their labour. Now, however, it's increasingly obvious that nature itself is being exploited and destroyed to such an extent that, in order to continue fuelling this 'juggernaut', the very basis of human civilization is being placed at risk: 'it is by no means certain that the essential conditions for the survival and development of civilized society as we know it today will continue to exist.' Essentially, the 'business-as-usual' capitalist model has now created a situation which is characterized by: 'a radical (and growing) disjunction between on the one hand the demands placed on the environment by the modern global economy, and on the other the capacity of the natural forces embedded in the environment to meet these demands.'[4]

The madness of *Stern*

Just how destructively 'out of control' fossil-fuelled capitalism is, was shown by the *Stern Review* in 2007. This was a massive study for the British government, undertaken by Nicholas Stern – a former chief economist with the World Bank. According to Foster, this was seen as 'the most important and most progressive mainstream treatment of the economics of global warming.' It focussed on the six greenhouse gases, as identified in 1997 in the Kyoto Protocol – rather than just CO_2 – which extended the 1992 UN Framework Convention on Climate Change, and attempted to establish what level of GHG concentrations would keep increases in the average global temperature 'at no more than 3C (5.4F) over pre-industrial levels.' According to the *Stern Review*, that could be achieved by stabilizing GHG concentrations at 550ppm – 'roughly double pre-industrial levels.'[5]

However, by the early twenty-first century, such combined GHG concentrations (CO_2e) were already at 430ppm. By 2010, most climatologists believed we needed to prevent the average global temperature from increasing by more than 2°C above pre-industrial levels – and that that would mean keeping combined GHG concentrations at 450ppm. Beyond that, climate scientists pointed out, various positive feedbacks and tipping points could well be triggered, 'leading to an uncontrollable acceleration of climate change.' In fact, several climatologists, including James Hansen, argued that CO_2 concentrations alone needed to be reduced – to no more than 350ppm – if we wished to preserve the conditions on Earth that had allowed human civilization to develop some 10,000 years ago, and to which life on the planet is currently adapted. Because 'most climate models failed to consider "slow" climate feedback processes such as the disintegration of ice sheets and the release of greenhouse gases from soils and the tundra', most climatologists came to agree with Hansen that CO_2 levels needed to be brought down as quickly as possible to 350ppm, because the more such concentrations rose, 'the greater the threat of creating irreversible environmental changes with dire consequences.' Many climatologists thus saw the levels envisaged by the *Stern Review* as likely to result in 'environmental effects [that] would undoubtedly be absolutely calamitous.' Hence the establishment of the international climate campaigning movement, 350.org, which was founded in the US in 2008 by, among others, leading environmentalist Bill McKibben.[6]

The 'madness' underlying the *Stern Review* was further shown by the fact that it accepted there was probably a one-in-five chance that the world would experience global boiling of *more* than 3°C, *even* if GHG concentrations were held at 430ppm. Even worse, it accepted that, if GHG concentrations rose to 550ppm, the chances of going above 3°C of global warming would rise to 30-70 per cent – with even a 10 per cent chance of exceeding 5°C! Yet even a 3°C rise would take the average global temperature to a level last seen approximately three million years ago. But it gets still worse: the *Stern Review* warned that the increase they envisaged could also trigger 'a shut-down of the ocean's thermohaline circulation that warms Western Europe.' That eventuality would cause such an abrupt change in the climate that western Europe could be plunged into Siberian-like conditions. Furthermore, research by other climatologists showed that if the average global temperature rose by 3°C, there could be a 90 per cent drop in water flow in the Indus River system by the end of this century – water which hundreds of millions of people depend on for growing crops. While others have argued that, with GHG concentrations at 550ppm, there was a five per cent chance that the average global temperature could rise as high as 8°C – or even higher! As Foster has argued, to 'stabilize' GHG concentrations at the level envisaged by the *Stern Review*, 'could be disastrous for the earth, as well as for its people.'[7]

A year after the *Stern Review* was published, Olivier Godard published an assessment of its findings. One of his conclusions was that a doubling of pre-industrial levels GHG concentrations – from 280ppm to the 550ppm level, as envisaged by the *Stern Review* – would likely be reached at some point between 2030 and 2060. At such a level of GHG concentrations, he argued that 'there would be a 20 per cent chance that average temperature would be in excess of 5 degrees.' As he stated, a 5°C rise in the average global temperature would be unprecedented in human history, corresponding 'to the same kind of difference as between the present situation and the last ice age.'[8]

Even before Godard's assessment, the summer of 2007 had seen the Arctic lose – in just one week – an area almost twice the size of the UK. By then, some scientists were already warning of the risk of an ice-free Arctic as the result of passing tipping points: perhaps in as little as twenty years. The effect of such an eventuality would be a positive feedback – the 'albedo effect' – whereby an enormous reduction in the Earth's reflectivity would sharply increase global warming. At the same time, fears were already mounting that there could be a rapid disintegration of ice sheets in West Antarctica and Greenland, with sea-level rises that would threaten to flood islands and coastal regions. So, even as the *Stern Review* was published, there were clear signs that significant changes were needed to stave off climate and ecological catastrophe.

Despite all these obvious and enormous risks to people and planet, it was realized as early as 2009 that the *Stern Review*'s 'stabilizing' figure of GHG concentrations at 550ppm would be reached by 2050, or even 2035 – even if the amount GHG emissions didn't increase beyond the levels of 2007. Unbelievably, even the *Stern Review* accepted this was unrealistic: given the 'business-as-usual' model of modern industrial societies, GHG emissions would continue to increase significantly. There was even a risk that, in the second half of the twenty-first century, GHG concentration levels could possibly rise to 750ppm, resulting in an increase in the average global temperature of more than 4.3°C. To counter this, the *Stern Review* optimistically saw GHG emissions peaking in 2015, with thereafter a yearly one per cent *reduction,* in order to stabilize concentration levels at 550ppm – which it argued stood a 'significant chance' of keeping the global average temperature increase to 3°C. Significantly – as regards 'out-of-control' capitalism – the *Stern Review* warned *against* any radical reduction in GHG emissions. This was because the economic costs 'would be prohibitive, destabilizing capitalism itself. … consequently, in order to keep the treadmill of profit and production going the world needs to risk environmental Armageddon.'[9]

Yet, as pointed out by Foster, and others, 'the existing [capitalist] social system is guaranteed not to deliver' the far-reaching radical changes needed to stave off climate and ecological collapse. Essentially, what the *Stern Review* – and later economic plans and climate agreements – have attempted to do is come up with 'solutions' that try to reduce some of the impacts of capitalism on Earth's ecology 'without challenging the economic system that in its very workings produces the immense environmental problems we now face.' However, if the destructive 'logic' of capitalism – 'the necessity of continued, rapid growth in production and profits' – is not ultimately ended, all attempts at 'solving' the climate and ecological crises will always fall far short of what's actually required.[10]

Farewell Paris?

So, with most climatologists arguing that combined GHG concentration levels shouldn't go beyond 450ppm – in order to keep 'the global temperature increase at about 2°C above pre-industrial levels' – what has happened to GHG concentration levels since 2007? Well, despite COP21 agreeing in Paris, in 2015, to try to keep any increase in the average global temperature to 1.5°C, combined GHG concentrations in the atmosphere have actually increased. By the end of 2021, total GHG concentration levels had 'reached 472 parts per million CO_2 equivalents.' This was revealed in a report published by the European Environment Agency in February 2024. This concentration was within the 'peak level' range that successive COPs since Paris – Glasgow (2021), Sharm el-Sheikh (2022) and Dubai (2023) – have agreed should not be exceeded 'if — with a 67% likelihood and not allowing a temperature overshoot — the global temperature increase is to be limited to 1.5°C above pre-industrial levels.' However, if a 'temperature overshoot' *is* factored in, the report stated that the peak level could be exceeded before 2026 – and that 'the peak concentrations corresponding to a temperature increase of 2°C by 2100 could be exceeded before 2032.' The crucial part of the report points out that the 2021 GHG concentrations showed an *increase* of 50ppm since 2011 – this was 'about 192ppm more than in pre-industrial times.'[11]

In March 2023, a Synthesis Report of the IPCC's Sixth Assessment Report had this 'high-confidence' warning about the risks of 'overshoot', should global warming exceed 1.5°C: 'Overshoot entails adverse impacts, some irreversible, and additional risks for human and natural systems, all growing with the magnitude and duration of overshoot.' Particularly worrying was that the Report stated that only 'a small number of the most ambitious global modelled pathways limit global warming to 1.5°C by 2100.' The Report also pointed out that the adverse impacts that would occur during even a relatively-short period of overshoot would cause 'additional warming via feedback mechanisms, such as increased wildfires, mass mortality of trees, drying of peatlands, and permafrost thawing, weakening natural land carbon sinks and increasing releases of GHGs.' All that would make any return to 1.5°C 'more challenging.' However, the level and duration of overshoot would risk exposing 'ecosystems and societies ... to greater and more widespread changes in climatic impact-drivers.' Consequently, the risks 'to infrastructure, low-lying coastal settlements, and associated livelihoods' would be high. Particularly at risk of 'irreversible adverse impacts' would be certain ecosystems with low resilience, such as: 'polar, mountain, and coastal ecosystems', which would experience ice-sheet and glacier melt, and higher rises in sea levels. Finally, the Report concluded that 'the larger the overshoot, the more net negative CO_2 emissions would be needed to return to 1.5°C by 2100.'[12]

However, as revealed by a report published by the UN's World Meteorological Organization (WMO) in November 2023, increases in the level of GHG concentrations continued into 2022. This report showed that the amount of 'heat-trapping greenhouse gases in the atmosphere' in 2022 had 'once again reached a new record' and that there was 'no end in sight to the rising trend.' For the first time, 2022 saw CO_2 concentrations – seen as the most important of the greenhouse gases – reach a level 50 per cent higher than pre-industrial times; worryingly, this trend had then continued into 2023. As the 2023 Report stated: 'The last time the Earth experienced a comparable concentration of CO_2 was 3-5 million years ago, when the temperature was 2-3°C warmer and sea level was 10-20 meters higher than now.'

Additionally, during the same period, the concentrations of other greenhouse gases increased significantly, leading WMO Secretary-General Professor Petteri Taalas to say that 'Despite decades of warnings from the scientific community, … we are still heading in the wrong direction.' As a result, he issued this stark warning: 'The current level of greenhouse gas concentrations puts us on the pathway of an increase in temperatures well above the Paris Agreement targets by the end of this century.' Such an increase would result in even more extreme weather, including intense heat and rainfall events, ice melt, sea-level rise and ocean heat and acidification. As a result, the socio-economic and the environmental costs would be vastly greater; as Taalas concluded: 'We must reduce the consumption of fossil fuels as a matter of urgency.'[13]

Highway to climate hell

For several years before 2023, the average global temperature had already increased to the warmest it had ever been during the 12,000 years of the Holocene. As a result, and as pointed out by many writers, climate change had already shifted if not actually destroyed 'habitat zones for animals and plants and influenced the hydrologic cycle. Specific positive feedbacks have been set in motion, so that even if carbon dioxide emissions do not increase further, significant additional warming would still occur.' However, despite all the COP agreements, the 'greenwash' from fossil fuel companies about investing in renewable energy, and largely-speculative talk about carbon capture and storage (CCS), it's important to know that the major fossil fuel companies are still intent on extracting and burning the bulk of remaining oil and gas deposits. It was this behaviour which led to UN Secretary-General, António Guterres to make a strong attack on fossil fuel corporations and governments of high-emitting countries, when he introduced the Third IPCC Report in April 2022. He accused them of lying, and 'not just turning a blind eye' but of 'adding fuel to the flames', and categorically stated that: 'We are on a fast track to climate disaster.… This is not fiction or exaggeration. It is what science tells us will result from our current energy policies.' He stressed that the world was now 'on a pathway to global warming of more than double the 1.5°C limit agreed in Paris.' Given the results of such a temperature rise would be catastrophic, he warned that 'This is a climate emergency. Climate scientists warn that we are already perilously close to tipping points that could lead to cascading and irreversible climate impacts.' The existing known global fossil fuel resources, if burned, are sufficient to completely de-ice Antarctica. Yet the fossil fuel companies remain determined to extract and burn as much as they can. Leaving aside the other impacts of burning all those reserves – such as deadly severe heatwaves around the world – doing so would lead to a rise in sea levels of more than 58 metres![14]

The possibility for such large rises in sea levels had been established as early as 2015 by a simulation study, carried out by Ricarda Winkelmann and others, which showed that 'burning the currently attainable fossil fuel resources is sufficient to eliminate the [Antarctic] ice sheet.' Assuming cumulative fossil fuel emissions at 10,000 gigatonnes of carbon (GtC), the study calculated that Antarctica would become almost ice-free, 'with an average contribution to sea-level rise exceeding 3m per century during the first millennium.' That study concluded that 'unabated carbon emissions thus threaten the Antarctic Ice Sheet in its entirety with associated sea-level rise that far exceeds that of all other possible sources.' Stevie Smith's poem 'Not Waving but Drowning' comes to mind![15]

The risk of disappearing ice sheets was further highlighted by a report which appeared at the end of February 2024, nine years after the 2015 study referred to above. The February 2024 report, by Benji Jones, was based on research from the Climate Change Institute of the University of Maine, and showed that the North Atlantic Ocean was unusually warm – in fact, it was 2°F hotter than the average temperature 'over the last three decades.' This was 'warmer than any other time on record', with the Atlantic having 'started breaking heat temperature records' in March 2023 – leading scientists to state that such a temperature anomaly was 'deeply troubling.' This is because oceans tend to absorb up to 90 per cent of the global warming resulting from rising concentrations of greenhouse gases. The report also noted that there were other signs that 'the planet, as a whole' was heating up – and referred to a new study in the journal *Nature Climate Change*, published earlier in February 2024, which showed planet Earth had already 'surpassed 1.5°C of warming, a limit set by the 2015 Paris Agreement to avoid the worst impacts of climate change.' According to that paper – by Malcolm T. McCulloch and others – their research showed that 'global warming was already 1.7 ± 0.1°C above pre-industrial levels by 2020.' As this result was 0.5°C higher than IPCC estimates, the researchers saw global warming reaching 2°C by the late 2020s – 'nearly two decades earlier than expected.' That *Nature Climate Change* paper also stated that their research showed 'that intense heatwaves and associated extreme events may now be the new normal'; and concluded that 'global mean surface temperatures (GMSTs) have, or will soon exceed, the Paris Agreement of 2015 of holding GMSTs to "well below 2°C above pre-industrial levels."'[16]

Yet fossil fuel companies have continued to push ahead with their climate-wrecking activities, which is no doubt what prompted UN Secretary-General António Guterres, in June 2024, to call those fossil fuel corporations the 'Godfathers of climate chaos', and to accuse them of having 'shamelessly greenwashed' while, at the same time, actively trying to delay any meaningful climate action by funding advertising companies and lobbyists to persuade the public and politicians that such action isn't necessary. In addition, although the negative impacts of global boiling are making the lives of millions of people much more difficult, if not impossible, those same fossil fuel corporations – increasingly seen by many as the 'fossil fuel Mafia' – are still receiving £trillions in subsidies from ordinary taxpayers while, simultaneously, raking in record profits. Guterres went on to say that we were now at the stage where it was a case of '*We* the peoples versus the polluters and profiteers.' As Malm and Carton observed, by 2024, it was clear that 'Things were completely, infernally, demoniacally out of control: the classes ruling the planet seemed bent on burning it as fast as physically possible.' Guterres thus praised climate activists for continuing to push for action, telling them that 'You are on the right side of history. You speak for the majority. Keep it up; don't lose courage, don't lose hope', and that, united, 'we can win.' The bottom-line, though, is that without *serious* action to control rising GHG emissions, the fossil fuel companies will continue pushing the planet along a 'highway to climate hell [and] planetary destruction.'[17]

Any doubt that people and planet are indeed being pushed towards real climate hell has been surely been removed by the European Commission Copernicus Climate Change Service report which came out at the end of May 2024. This report – which in part led to Guterres' speech the following month – showed that May 2024 had been the hottest May in history. In order just to keep to the 1.5°C limit for temperature rise, GHG emissions need to drop by nine per cent *every* year – but, in 2023, they actually *rose* by one per cent. Sadly, this was just the continuation of a trend. As regards CO_2 alone, the European Environment Agency had, in February 2024, reported an increase in CO_2 emissions between 2012 and 2022: 'the annual average concentration of CO_2 reached 414ppm and 417ppm in 2021 and 2022 respectively (+137ppm or +148%) above pre-industrial levels.' In fact, on the same day that Guterres made his speech, the UN's WMO reported that there was now an 80 per cent chance that the 1.5°C limit would be passed before 2030. Thus, it should be clear that 'It's climate crunch time', and that the need for global action to tackle rising GHG emissions is now unprecedentedly urgent; or, as Guterres put it: 'we stand at the moment of truth.' According to him, we needed to fight harder, because all depends on decisions taken by political leaders in this current decade – but 'especially in the next 18 months.'[18]

Terror and despair in 2024

In his speech, Guterres referred to extreme weather events that, during the previous month, had destroyed lives and caused vast economic damage from east Asia to the west coast of the USA. Further confirmation of where we're heading was provided by a series of other reports that also came out in May 2024. A report in *The Guardian* showed that many of the world's top climate scientists were now 'terrified' – and in despair – because of the failure to halt increases in GHG emissions. Climate scientist Dr Ruth Cerezo-Mota stated that she now expected 'the world to heat by a catastrophic 3°C this century, soaring past the internationally agreed 1.5°C target and delivering enormous suffering to billions of people.' Worryingly, she saw '3°C is being hopeful and conservative'; although she made it clear that even 1.5°C was bad, she didn't think there was any chance of sticking to that 'limit': 'There is not any clear sign from any government that we are actually going to stay under 1.5°C.' Such pessimism emerged from a survey of the world's leading climate experts, which showed that 77 per cent now believed that average global temperatures will reach a 'devastating degree of heating' of 'at least 2.5°C above preindustrial levels'; while 42 per cent believed the increase will be more than 3°C. Despite this gloom and doom, most climate scientists are still determined to keep reporting the facts so that, even if they fail to persuade corporations and governments to take the necessary actions, at least none of those guilty of driving us into a 'climate hell' will be able to claim that they did not know. The bottom line for those scientists was that the dangers facing humanity – if *serious* action is not taken – are so devastating, that many of them envisage 'a "semi-dystopian" future, with famines, conflicts and mass migration, driven by heatwaves, wildfires, floods and storms of an intensity and frequency far beyond those that have already struck.' Some – like Camille Parmesan, from the CNRS ecology centre in France – said that, at one point, she was thinking of giving up because of the failure to stop the increase in GHG emissions. But was inspired to keep going by the enthusiasm displayed by young climate activists. This determination of climate and environmental activists to keep struggling has also helped Cerezo-Mota – despite the lack of serious actions by those with the power and money to make the necessary changes – to cope with her fears and despair: 'then I see the younger generations fighting and I get a bit of hope again.' Another scientist, Henri Waisman of the IDDRI policy research institute in France, revealed that he felt 'moments of despair and guilt' for not having been able to bring about the the changes that were needed – such feelings had increased since he'd become a father. As well as seeing some progress since 2005, he was also helped by the thought 'that every tenth of a degree matters a lot – this means it is still useful to continue the fight.'[19]

One explanation of why so many of those climate scientists despaired, and believed we were indeed on a 'highway to climate hell', had emerged in April 2024, when a report from Carbon Brief had argued that 2024 was already 'shaping up to either match or surpass 2023 as the hottest year on record.' The report pointed out that, for the first three months of 2024, global temperatures had been 'exceptionally high' – at a level that was 'around 1.6°C above pre-industrial levels.' In fact, the start of 2024 saw record-high global temperatures across vast swathes of the planet, including in the tropical Atlantic and western Pacific oceans, much of South America, Central Africa, the Mediterranean and the Indian Ocean. Carbon Brief fully expected that April 2024 would also set a record high, thus making it the eleventh month in a row whereby each month set new temperature records. Their prediction was that, based on 'normal' temperature patterns, 2024 was 'virtually certain to be either the warmest or second-warmest year on record.' However, what was particularly worrying was that, because 2023 had significantly broken the 'normal' climate patterns, there was an element of uncertainty over predicting what would happen in 2024. The report thus stated that 'If the latter half of 2024 ends up similar to 2023, there is a worry that we might be entering what has been described as 'uncharted territory' for the climate.'[20]

Sadly, there was even more evidence in May 2024 that we are indeed rushing towards a 'climate hell.' In that same month, another report revealed that, just like land temperatures, the world's oceans had also 'broken temperature records every single day over the past year', and that nearly 50 days had 'smashed existing highs for the time of year by the largest margin in the satellite era.' Those figures were based on those from the EU's Copernicus Climate Service and, as the report pointed out, for many decades, 'the world's oceans have been the Earth's "get-out-of-jail card" when it comes to climate change.' This is worrying because not only do the oceans absorb around 25 per cent of CO_2 emissions but, as noted earlier, they also soak up around 90 per cent of excess heat. Evidence showed that, over the past year, the oceans – especially as regards sea surfaces – were clearly struggling to cope with increasing temperatures. In particular, the figures showed that not only had 'every single day since 4 May 2023 broken the daily record for the time of year', and that on some days the margin had been huge. According to Professor Mike Meredith of the British Antarctic Survey: 'The fact that all this heat is going into the ocean, and in fact, it's warming in some respects even more rapidly than we thought it would, is a cause for great concern.' He saw all that as real signs of the environment moving into 'undesirable' states, and stated that 'if it carries on in that direction the consequences will be severe.'[21]

Also in May 2024, another report made it clear how even a 1.5°C increase posed serious problems, including more intense heatwaves and storms; the possible triggering of tipping points for ice-sheet collapses, and the thawing of permafrost; and the die-off of tropical corals. Importantly though, climate scientists didn't see 1.5°C as 'a cliff-edge leading to a significant change in climate damage.' Instead, because the climate crisis was still increasingly incrementally, 'every tonne of CO_2 avoided' would reduce people's suffering. However, at 2°C, the global increases in direct flood damage would double, while the chances of brutal heatwaves like the one that hit the Pacific north-west in 2021 would be '100-200 times more likely.' At 2.7°C, 'two billion people would be pushed outside humanity's "climate niche"' – the stable and benign circumstances that have allowed human civilizations to emerge over the past 10,000 years; while at 3°C, cities like Shanghai, Rio de Janeiro, The Hague and Miami would 'end up below sea level.' If the rise in the average global temperature were to exceed 3°C, 'the impact of climate shocks in one place will cascade around the world, through food price spikes, food and water shortages, broken supply chains, and refugees by the millions.'[22]

Rising temperatures – increasing instability

Yet there's still more: June 2024 became the thirteenth consecutive month for record temperatures. According to a report that month from the EU's Copernicus Climate Service, the average global temperature for the previous twelve months (July 2023 – June 2024), had been 'the highest on record, at 0.76°C above the 1991-2020 average and 1.64°C above the 1850-1900 pre-industrial average.' Sea surface temperatures also continued to rise and break records, leading Carlo Buontempo, the Director of Copernicus to state that: 'Even if this specific streak of extremes ends at some point, we are bound to see new records being broken as the climate continues to warm. This is inevitable, unless we stop adding GHG into the atmosphere and the oceans.' While, according to a report from the World Weather Attribution (WWA) group at the end of May, the seemingly 'never-ending' rain that hit the UK and Ireland in the autumn and winter of 2023 was made '10 times more likely and 20% wetter' because of global boiling, with the period October 2023 to March 2024 being 'the second-wettest such period in nearly two centuries of records.' For those who may have forgotten, the results of that heavy and prolonged rainfall were 'severe floods, at least 20 deaths, severe damage to homes and infrastructure, power blackouts, travel cancellations, and heavy losses of crops and livestock.' A separate analysis by the Energy and Climate Intelligence Unit estimated to cost to farmers would be around £1.2bn. Those warnings were borne out at the end of September 2024, with many parts of Wales and central and southern England suffering severe flooding after torrential rainfall, with over one month's rain falling in just 48 hours. According to the scientists involved, the amount of rain caused by storms such as Babet, Ciarán, Henk and Isha – which, before the climate crisis, would normally have occurred around once in fifty years, could now be 'expected every five years owing to 1.2°C of global heating reached in recent years.' Their conclusion was that, if fossil fuel burning is not rapidly cut, so that the average global temperature rose by 2°C in the next decade or two, 'such severe wet weather would occur every three years on average.' This is just more evidence of how global boiling is super-charging extreme weather events – such as heatwaves, wildfires, droughts and storms – across the world. Dr Sarah Kew – a researcher at the Royal Netherlands Meteorological Institute and a member of the WWA team – stated that: 'The UK and Ireland face a wetter, damper and mouldier future due to climate change.' According to her, unless the world reduces GHG emissions to net zero, 'the climate will continue to warm, and rainfall in the UK and Ireland will continue to get heavier.' Those warnings were later borne out at the end of September 2024, with many parts of Wales, and central and southern England, suffering severe flooding after torrential rainfall, with over one month's rain falling in just 48 hours. In fact, September 2024 turned out to be the UK's wettest September since records began.[23]

Also, at the end of May 2024, it was reported that India's capital, Delhi, had recorded its hottest day ever, with temperatures reaching just over 49°C during what was described as a 'scorching heatwave.' Apart from the risks from heat exhaustion – especially for children, elderly people and those with chronic diseases – the authorities also warned that water shortages were likely. That heatwave was just one more example of how 'the climate crisis is causing heatwaves to become longer, more frequent and more intense.' The heatwave had been building up inexorably for weeks, but even so residents were shocked by how extreme the conditions were: 'People told of fingers being scorched from touching the steering wheel of a car, and tap water was coming out at boiling temperatures.' However, temperatures reached even higher in Jaipur, the capital of Rajasthan, peaking at 50.5°C; according to one hospital, so many casualties of the heat arrived at the mortuary that its capacity had been exceeded. Many of the victims were 'poor labourers, who have no choice but to work outside, and homeless people' – yet another clear reminder that the main victims of the climate crisis are working class and poor people. In the same month, Cyclone Remal hit India and Bangladesh, killing thirty-eight people – with the Bangladesh Meteorological Department saying 'the cyclone was "one of longest in the country's history" and [blaming] climate breakdown for the shift.' At end of August 2024, Bangladesh was once again hit by an extreme weather event – this time, by a devastating flood which saw over 1.2 million families trapped in floodwaters which swept across the country. This hit over 5.8 million people across eleven districts, with people losing homes, belongings and livelihoods. Crops and freshwater were also destroyed, along with roads and railways – the latter hindering evacuation, medical services, and relief. According to the UN's Office for the Coordination of Humanitarian Affairs, the costs to agriculture were enormous: 'A total of 296,852 hectares of crops have been affected by the flood. Loss of fisheries is USD 122 million and livestock loss is USD 34 million initially.'[24]

Such recent developments have merely underlined the warnings about the dangers of passing tipping points pointed out, in January 2023, by leading climate scientist Johan Rockström, in a speech delivered for the World Economic Forum at Davos. In it, Rockström – the Director of the Potsdam Institute for Climate Impact Research – warned of how passing just a few of sixteen tipping points could push our entire planet into crisis. As he pointed out, nine of those were already showing significant 'signs of instability' – 'push them too far too far, and they will shift over from supporting humanity to starting to undermine humanity.' Of those nine tipping points, four were now showing signs of 'being at risk already at 1.5°C.' Of those, the West Antarctic ice sheet and the Greenland ice sheet would – if both disintegrated as a result of continued global boiling – result in '10 metres sea level rise.' His conclusion was that: 'This shows scientifically that 1.5°C is a physical limit. It is not a political target.' However, he went on to present even greater dangers: 'We have more and more scientific evidence of connectivity between these tipping elements. Cascade risk of dominoes.' As he pointed out: 'This is not a climate crisis. We are now facing something deeper.' With mounting evidence of mass extinction, increasing air pollution, and the undermining of ecosystem-functioning – all now really putting humanity's future at risk – his warning was stark: 'This is a *planetary* crisis.' Less than a year later, in November 2023, James Hansen was lead author of an article in the *Oxford Open Climate Change* journal, which predicted that, if GHG emissions continued at the current rate, 'global warming will exceed 1.5°C in the 2020s and 2°C before 2050.' Particularly worrying was that the paper pointed out how, because there is a 'delayed response' as regards both rising CO_2 emissions and 'amplifying feedbacks', the full impact of current GHG emissions might not be seen for at least several decades – thus resulting in 'intergenerational injustice.' Later, in July 2024, in a longer TED talk in Seattle, he said this: 'What worries us most is this: we are starting to see an acceleration of warming over the past 50 years.' If that trend continues, he warned that 'we will crash through two degrees Celsius within 20 years, and hit three degrees Celsius by the year 2100.'[25]

As if to confirm Rockström's concerns, a report in early October 2024 warned that, because of wildfires, forests around the world were being 'changed from carbon sinks into carbon sources', thus making it harder to 'slow global heating.' In particular, the authors of this study – which was led by the UK Met Office – concluded that wildfires were burning through the 'carbon budget' that had previously been calculated as allowing global heating of the planet to be limited to a safe level. In fact, according to them, there was an 'accelerating trend' which was 'approaching – and may have already breached – a "critical temperature threshold" after which fires cause significant shifts in tree cover and carbon storage.' According to the Met Office, such impacts were now being triggered at an average global temperature just 1.34°C above pre-industrial levels. The last two years – the Earth's two hottest years in recorded history – have seen forests 'going up in smoke in Brazil, the US, Greece, Portugal and even the Arctic Circle.' Such wildfires were having a 'double impact on the global climate': emitting CO_2 from the burned trees, and then reducing the capacity of forests to absorb CO_2. Other research showed that Latin America is 'becoming warmer, drier, and more flammable', with the Amazon now undergoing 'a "critical slowing down", with more than a third of the rainforest struggling to recover from drought', after four 'one-in-a-century' prolonged dry spells in less than twenty years.[26]

Teetering on a planetary tightrope
However, the evidence of capitalism's ecocidal direction continued to mount. The end of October 2024 saw further extreme weather events and climate reports – all confirming current trends. In eastern India, the severe threats posed by Cyclone Dana saw the authorities forced to evacuate 1.5 million people from its expected path; while in the northern Philippines, Tropical Storm Trami (known locally as 'Kristine') 'dumped one month's rain over 24 hours', leaving more than 20 people dead, and forcing over 150,000 to evacuate. Then, at the very end of October 2024, Taiwan experienced Typhoon Kong-rey – as reported by the BBC, 'recent years have seen typhoons with stronger, more destructive winds and heavier rains.' While in the global North, an 'extreme weather event' hit the Valencia region of Spain (which produces 'two-thirds of the citrus fuit grown in Spain') on 29 October. There, over a year's worth of rain fell in eight hours, causing unprecedented heavy and destructive flooding: by the beginning of November, the death-toll had topped 220, with more deaths expected once the clear-up operation ends. Those torrential rains also hit the Huelva region, while the city of Cartaya had two months' rain in ten hours. That same week, there were also devastating floods in Santa Marta (Colombia), Halabja (Iraq), and sukabumi (Indonesia). A later article in *The Guardian* called Valencia's floods 'apocalpytic', and claimed they showed 'two undeniable truths: the climate crisis is getting worse and Big Oil is killing us.' Though, of course, it's capitalism as a whole, rather than the fossil fuel industry, which is ultimately to blame for all these climate and ecological disasters. As if to explain the increasing number and severity of such extreme weather events, the UN issued its annual 'Emissions Gap Report', which warned that the climate policies currently being followed by the world's governments – which

had seen global annual emissions continue to increase – 'could lead to 3.1°C warming by 2100.' Calling for *greater* action on 'slashing planet-warming emissions', UN Secretary-General António Guterres issued one of his starkest warnings: 'We're teetering on a planetary tight rope. Either leaders bridge the emissions gap, or we plunge headlong into climate disaster.'[27]

Increasingly, climate experts believe that the 1.5°C limit is now 'lost', *regardless* of what's decided at COP29 in Azerbaijan in November 2024. In addition, using their new 'Climate Pulse' tool for monitoring the state of the world's climate the Copernicus Climate Change Service (C3S) issued a report which confirmed that the average global temperature is now higher than that set in 2023, and now stands at 1.7°C higher than pre-industrial times, with GHG levels still increasing, producing yet more temperature 'anomalies' as regards air and ocean temperatures, and with both poles now warming faster than ever. This situation was confirmed, in early November, by a further report in *The Guardian*, which showed that the European Union's space programme now considered it 'virtually certain' that 2024 will indeed be the hottest year on record, with the average global temperature exceeding the 1850-1900 average by 1.62°C. According to Dr Samantha Burgess, Deputy Director of C3S, stating that 'This marks a milestone in global temperature records', which should act as a 'catalyst to raise ambition' at the forthcoming COP29 meeting. Such findings were largely confirmed in November 2024 – as COP29 was getting underway – when the WMO's 'Update' revealed that 2023 had indeed been the hottest year on record, and predicted that 2024 was set to be even hotter. Furthermore, the 'Update' also stated that the period 2015-24 would be 'the hottest decade on record.' With Trump – who's on record as describing climate change as a 'hoax' – now US president-elect and urging 'Drill, baby – drill!'; and the president of COP29 caught trying to negotiate new fossil fuel deals, the prognosis for this year's COP in Azerbaijan was already not very encouraging. Only days later, anger broke out at the meeting when Saudi Arabia announced it wouldn't sign an agreement that committed to reducing fossil fuel use. This would clearly undermine the COP28 agreement which saw 'consensus' on the need to transition away from fossil fuels – not surprisingly, Saudi Arabia's announcement led to several countries protesting. Ed Miliband, for the UK, said: 'Standing still is retreat and the world will rightly judge us very harshly if this is the outcome.' If Saudi Arabia's stance scuppers any meaningful outcome for COP29, this could prove to be the proverbial 'writing on the wall' for the entire COP process.[28]

Having thus established just how 'out of control' the capitalist system now is, the next chapter will attempt to explain *why* things have developed – and are continuing to develop – in ways that are increasingly, and fundamentally, destroying the ecological sustainability of planet Earth.

DEADLY PARADOXES AND KILLER CORPORATIONS

States, corporations and financial institutions aren't as separate as you might think. Most of the time these institutions ... [are] working with and through each other. ...modern states increasingly see their priorities as similar to those of corporations; ...These similarities – forged through neoliberal rule – are the foundations of state-corporate cooperation.[1]
Grace Blakeley

Having seen just how 'out of control' capitalism has now become, it's necessary to look closer at reasons which explain why it continues to pursue its increasingly-ecocidal activities across the globe. Essentially, capitalism has, at its core, some deadly paradoxes which are part of its DNA. In addition, we need to examine the very nature of capitalist corporations themselves; and how those corporations have 'captured' states around the world to ensure their operations continue – regardless of the human and planetary costs.

Paradoxes of capitalism

One of the reasons why fossil fuel companies, and all those other capitalist corporations addicted to constantly increasing GDP and capital accumulation – along with their political supporters and enablers – persist in ignoring the mounting scientific evidence of the dangers posed by global boiling and ecological destruction is because many believe that 'some day', 'someone' will eventually come up with a technological 'fix' to capitalism's destructive impacts. This means therefore that they don't have to change what they're doing now and so can, instead, continue with 'business-as-usual.' Yet history has shown – even from the early days of capitalism's Industrial Revolution – that this is essentially overly-optimistic and unfounded dreaming.

More specifically, there are two paradoxes which highlight both the irrationality and the contradictions of capitalism. Under a 'natural economy', efficiencies in energy-use would not necessarily result in increased production, as once a sufficient level of use-values had been reached, any extra time could be devoted to other aspects of living. But, under capitalism, there is an inherent drive to make use of such efficiency improvements to produce even more – in order to *constantly* increase capital accumulation for owners of capital. In addition, there is the stark contradiction that, under capitalism, any increase in private wealth for the minority property-owning class almost always comes at the expense of *public* wealth – an aspect of capitalism's robbery of both people and planet.

The Jevons Paradox

The first of these two paradoxes was pointed out by the English political economist, William Stanley Jevons. In 1865, in his study *The Coal Question*, he convincingly showed that 'improved efficiency in the use of coal made it more cost effective as an energy source.' As a result, there was a rise, not a decline, in the extraction and use of coal – and therefore a rise in CO_2 emissions. This discovery, which established a connection between greater efficiency, and the subsequent greater use of coal – a phenomenon described by John Bellamy Foster as 'a dilemma for ecological modernization' – became known as the *Jevons Paradox*. Yet, despite making this connection, Jevons didn't explicitly link this to capitalism's constant pursuit of ever-greater production and profits. In large part, this was because he saw capitalism 'more as a natural phenomenon than a socially constructed reality', and believed that 'the resources of nature are almost unbounded.' However, as others showed later, production under capitalism is essentially about greater profits and the accumulation of private wealth through constant economic expansion – *not* about satisfying humanity's social needs or for ensuring environmental sustainability.[2]

Thus the 'logic' of capitalism will always be about ever-greater extraction and use of the Earth's resources, and the consequent increase in waste and pollution, which together threaten to undermine and even destroy nature's ecosystems. This is why any efficiency improvements, under capitalism, will always be overtaken by increased economic growth and expansion. This argument has been confirmed by a study of several advanced capitalist states over the period 1975-96. While the carbon efficiency of the US, Japan, The Netherlands and Austria increased dramatically in that period, 'total carbon dioxide emissions, and even per capita emissions, increased in all four of these nations.' A later study found the same results for Germany. Yet capitalists – and supporters of their system – continue to 'assert that capitalist development will lead to improved technologies and efficient raw material usage, and that this will decrease emissions and environmental degradation.' This is why those who argue that a 'green capitalism' is possible, believe that – if a realistic 'value' were put on nature and its 'free' services – 'capitalism would develop in an ecologically benign direction.' According to them, capitalist modernization will eventually lead to 'the decoupling of the economy from energy and material consumption, allowing human society, under capitalism, to transcend the environmental crisis.' Yet, while technological improvements have led to *relatively* greater fuel efficiency in cars, this has not led to a drop in the *absolute* number of miles driven, or the total consumption of fuel – instead, the number and size of vehicles on the roads have just increased. While, in other areas of economic activity, energy-efficiencies simply lower costs and increase sales. Yet some still claim that capitalism can 'incorporate an interest in the common good of society into its operations.' As has been seen – and will be seen later, when considering another of capitalism's paradoxes – history has repeatedly shown this is not the case.[3]

In fact, it is the countries which are generally more energy-efficient which are also 'the biggest consumers of natural resources', and the greatest emitters of greenhouse gases – simply because of capitalism's continual expansion of commodity production in order to accumulate ever-more capital. In prioritizing capital accumulation over social needs and ecological sustainability, capitalism also 'generates enormous waste throughout its operations. … in concentrations that threaten ecosystems.' As has been established by several researchers, capitalist 'economic growth and expansion typically outstrip gains made in efficiency.' Marx was in fact one of the first to note that capitalism, because of its continual drive for increased accumulation of capital, was unable to achieve a 'truly rational application of new science and technologies.' That is why such policies as carbon-markets and technologies such as carbon-capture have so far failed to achieve any significant reductions in GHG emissions or deal with global climate change. Modern-day researchers – such as the environmental sociologist Stephen Bunker – have found that, over a long historical period, 'the world economy as a whole showed substantial improvements in resource efficiency…, but that the total resource consumption of the global economy continually escalated.' As with capitalism's failure, over more than one hundred years, to significantly recover agricultural nutrients for the soil, its limited attempts to 'recover' CO_2 from the atmosphere are likely to prove at least as unsuccessful – if not more so. Plus, of course, in the meantime, capitalist expansion just keeps adding more greenhouse gases to the atmosphere, and dumping more waste on the natural world. Thus, in the twenty-first century, the Jevons Paradox, which has proved to be integral to the 'regime of capital', has returned to haunt capitalism and its supporters – and *all* Earthlings on planet Earth – with a vengeance: 'an economic system devoted to profits, accumulation, and economic expansion without end will tend to use any efficiency gains or cost reductions to expand the overall scale of production.' This is why the most affluent nations, as well as displaying greater eco-efficiency, also have a higher per capita ecological footprint. Consequently, it should now be abundantly clear that what people and planet need, as soon as possible, is an economic and social system 'driven by human development – which by necessity must also be ecological to sustain the conditions of life – not the accumulation of capital.'[4]

Currently, fossil fuel capitalists – and their ecological modernization defenders – are trying to convince us that we don't need to worry about increasing levels of CO_2 because they have 'found' a technological solution: 'carbon capture and storage' (CCS). This, they argue, will mean production and consumption levels can be maintained, as the 'new' technology will suck out and store excess greenhouse gases from the atmosphere. Shell, for instance, claim they will have 2,500 large-scale CCS facilities operational by 2050, with more than 10,000 such facilities by 2070. Exxon are more 'cautious', and predict somewhere between 250 and 500 CCS facilities will be operational by 2050.

However, researchers have also found that, elsewhere, Exxon have – worryingly – said that '"global scale is limited" for CCS and hydrogen tech by 2050.' Significantly, the IPCC have said that even if 'realized at its full announced potential, CCS would only account for about 2.4 per cent of the world's carbon mitigation by 2030.' While the Institute for Energy Economics and Financial Analysis (IEEFA), a non-profit think tank that carries out market-based research on the 'energy transition', has revealed that 'not one single CCS project has ever reached its target CO_2 capture rate.' Another researcher, Mark Jacobson, of Stanford University, has pointed out that because CCS itself also requires energy and materials to operate, 'CCS attached to a fossil-fueled power plant is still worse for the climate than replacing fossil energy with renewables.' According to research he undertook in 2019, CCS facilities 'actually increase carbon dioxide emissions by doing this, in addition to increasing air pollution' – even if powered by wind! His conclusion was that, from a climate perspective, CCS is not worth doing: 'If you just used wind to replace coal in the first place, you'd get a higher reduction in CO2 emissions.' In addition, proponents of CCS tend not to mention that there are long-term safety issues with underground storage of CO_2. Earthquakes could cause leakage, while some geologists and geo-physicists fear storage in geological strata could itself trigger earthquakes or volcanic eruptions and thus possible tsunamis: 'CCS could therefore pose a danger to future generations in the same way as nuclear waste.' Thus, dealing just with fossil fuels – even factoring in unsubstantiated hopes about CCS – it seems fair to say we need to end the fossil fuel era as soon as possible, before the fossil fuel era ends our chances for a liveable future on Earth.[5]

The Lauderdale Paradox

As well as the Jevons Paradox, another paradox also helps explain capitalism's 'destructive uncontrollability' – this is *Lauderdale Paradox*. In the history of economics, the Lauderdale Paradox provides an understanding of why capitalism's socially-destructive actions and its ecological contradictions will remain essentially 'out of control' as long as capitalism itself continues to operate. In 1804, the Earl of Lauderdale – James Maitland – published a book which argued that there was 'an inverse correlation between public wealth and private riches such that an increase in the latter often served to diminish the former.' Thus, under capitalism, the 'creation' of wealth is, paradoxically, achieved via the creation of scarcity – or what has been described as 'managed" scarcity.' Essentially, there is 'natural scarcity and capitalist-manufactured scarcity.' Aspects of this were covered in Chapter 6, when early capitalism deliberately deprived the majority of people of access to the commons and commons rights, in order to create food-scarcity that would force the newly-deprived to work on their large estates and, later, in their factories.[6]

According to Lauderdale, 'public wealth' consisted of useful things needed by people, whereas 'private riches' included the added element of existing 'in a degree of scarcity.' Essentially, the former were 'use values' – whereas the latter needed scarcity to have 'exchange values' and so augment *private* wealth. Thus, if necessities such as food or water became scarce, exchange values would be attached to them, thus increasing the private riches of those who possessed food or water – or who controlled access to them. The important thing to note is that, while enriching certain individuals, this would be 'at the expense of the common wealth.' Lauderdale assumed that people would revolt at the idea of a few individuals enriching themselves 'by creating a scarcity of any commodity generally useful and necessary to man.' Yet he was also aware that the – capitalist – society he lived in was 'already, in many ways, doing something of the very sort.' Capitalism's increasing search for private riches, based on establishing monopolies, was thus at the expense of 'the public wealth of society and the commons.'[7]

That this robbery was 'intrinsic to capitalist production' was noted by Marx, who used Lauderdale's explanation of an inverse ratio between use value and exchange value to develop aspects of his own critique of capitalism. In particular, Marx explained how the contradictions between use value and exchange value underpinned how, under capitalism, 'nature was rapaciously mined for the sake of exchange value': 'The earth is the reservoir, from whose bowels the use-values are to be torn.' It was this attitude to nature on the one hand, and private profit and capital accumulation on the other, that led to his understanding of how capitalism created an increasingly-deadly 'metabolic rift' between humans and the rest of the natural world. Marx was particularly critical of how capitalism created big private monopolies in land which thus 'divorced the majority of humanity' from access to food production and 'a communal relation to the earth.' It was capitalism's robbery of the commons – and of nature itself – which led Marx to argue that capitalism created such a fatal 'rift in the human-earth metabolism' that 'the reproduction of natural conditions was undermined.' His conclusion was that capitalism was 'an unsustainable system of production... rooted in the unceasing exploitation and pillage of human and natural agents.'[8]

For Marx – unlike for capitalism – nature was key to human existence and general social wealth, and labour was itself a part of nature (as well as being another source of social wealth). All useful products were 'combinations of two elements, the material provided by nature, and labour. ... Labour is therefore not the only source of material wealth.' Marx actually quoted William Petty (the founder of classical political economy) to emphasise this point: 'labour is the father of material wealth, the earth is its mother.' It is capitalism's failure to take nature into account, and 'its tendency to confuse value with wealth' which explains the 'fundamental contradictions of the regime of capital itself.' It is such fundamentally destructive contradictions that explain why capitalism, as a system, only thinks short-term and thus prioritizes accumulation of capital at the expense of real wealth which is, ultimately, rooted in the natural world. Marx summed up capitalism's attitude to the long-term conditions of people and planet thus: *Après moi le deluge!*' (roughly meaning: 'after me – disaster!'). As long as profits are being made now, what happens in the future is relatively unimportant.[9]

Modern-day ecological economists still see the value of the Lauderdale Paradox as helping to explain capitalism's 'repeated failures to apprehend the extent of the dangers facing us', and the 'narrow accumulation strategies' put forward by 'orthodox' economics as possible solutions. For instance, one study concluded – somewhat bizarrely! – that, because agriculture only accounted for some three per cent of the USA's national output, the failure of the agricultural sector, because of climate impacts, 'would have little impact on the country as a whole.' Another concluded that: 'global warming under business-as-usual would have a "negligible" effect on world output'; while yet another felt able to take comfort from his conclusion that even 'if agricultural productivity were drastically reduced by climate change, the cost of living would [only] rise by one or two per cent.' That such comments have been made at a time when at least one billion people across the globe are already experiencing serious food shortages, graphically illustrates the irrational and inhumane 'destructive uncontrollability' of modern-day capitalism. As Foster and Clark state: 'There is no understanding ... of production as a system, involving nature (and humanity), outside of national income accounting.'[10]

As regards food alone, 'nothing is more dangerous to the capitalist system than abundance. Waste and destruction are therefore rational for the system.' Such contradictions and paradoxes suggest that it cannot be assumed that even rising ecological costs 'can be counted on to check capitalism's destruction of the biospheric conditions of civilization and life itself.' As Naomi Klein has established, capitalism actually sees disasters of all kinds as yet more opportunities to make a profit. In fact, any scarcities arising from climate change chaos are likely to be grabbed by capitalism as an opportunity to further privatize the world's commons – even though this 'only accelerates the destruction of the natural environment, while enlarging the system that weighs upon it.' The using up and pollution of freshwater sources – mainly by intensive capitalist animal 'agriculture' – is just one example. As early as 1998, the UN Commission on Sustainable Development proposed that governments should approach '"large multinational corporations" in addressing issues of water scarcity' and even 'establishing "open markets" in water rights.' Large corporations are already positioning themselves to profit from water shortages; and the same is happening with the shift to fuel crops, with fears that fossil fuels will soon become exhausted. The result of this is already obvious in many global South countries, where large firms have replaced food crops with agro-fuels: resulting in greater food scarcity, rising food prices and increasing hunger. A system that sees profit opportunities in world hunger, and even in the increasing destruction of the natural world, is surely a system that needs to be consigned to the dustbin of history as quickly as possible. Thus, Lauderdale's Paradox underlines 'the supreme contradiction of a system that sees nature as a mere means of accumulation' and that even seeks 'to profit from the destruction of the planet.'[11]

The killer corporation

In October 2017, it was revealed that oil giants BP (originally known as the First Exploitation Company!) and Shell were both privately asking their scientists to come up with ways which would enable them to continue their operations – even if the average global temperature were to rise, by the middle of this century, to catastrophic levels as high as 4°C, or even 5°C, above pre-industrial times. This was while those companies were, at the same time, publicly claiming to support the 2015 Paris climate agreement to keep such increases to below 2°C and, if possible, to limit any rise to 1.5°C. This all came to light because of two reports published by the campaign group, Share Action. As was noted in an article in *The Independent*, 'The discrepancy demonstrates that the companies are keeping shareholders in the dark about the risks posed to their businesses by climate change.' As was known at the time, many climate scientists were already warning that even a 2°C rise would be 'catastrophic for the planet', as it would lead to the Arctic Ocean becoming totally ice-free for the first time in 100,000 years. This would have serious and adverse impacts on 'the weather in much of the northern hemisphere and speed up global warming.' Even before those revelations about BP and Shell's totally-irresponsible intentions, scientists had already been warning that the world was actually heading towards 'anything from 2.6°C to 3.1°C by the end of the century.' Yet we now know that from the late 1970s and early 1980s, both Shell and Exxon had been informed by their scientists that CO_2 emissions from fossil fuels would cause devastating global warming by as early as the mid-twenty-first century – if extraction and use of fossil fuels continued at the levels then current. Their first response was to conceal the evidence from society at large – and then, when independent climate scientists showed what was happening – they spent $bns to deny climate change, and then to limit effective action.[12]

To gain some idea of what a world heated to 5°C above pre-industrial levels would be like – something it seems BP and Shell are prepared to risk creating – it's worth looking at what Mark Lynas concluded in 2020: 'At five degrees, humanity has lost control of global temperatures, which are now spiralling relentlessly upwards. Food production is decimated, and large areas of the planet are too hot for humans to inhabit.' An increase in average global temperatures to 5°C above pre-industrial levels, would see Earth hotter than it has been for 55 million years. As a result, all ice sheets would disappear, thus reducing Earth's reflectivity and so accelerating global boiling. In addition, 'the thawing of the tundra will release massive quantities of the potent greenhouse gas methane'; while the resulting drought conditions would see the loss of the Amazon rainforest and thus its ability to sequester CO_2. There would, as a consequence of all that, be a mass wipe-out of life – leading to a situation that could be called, 'without hyperbole, threatened apocalypse.' Or, as Lynas warns, at that level of global boiling, 'It is almost game over.'[13]

So, what *are* the forces driving the climate and ecological crises – and increasingly pushing people and planet to the cliff edge? In particular, why does the 'logic' of capital accumulation end up being so diametrically opposed to environmental sustainability? Why does the operation of the capitalist economy threaten to undermine the entire atmosphere of the planet? According to some researchers, the answers to such questions can be found, to a very large extent, in the very nature and functioning of capitalist corporations. Once 'incorporated' in law, a capitalist corporation gains a legal status – separate from its shareholders or investors – known as 'corporate personhood', and can only be ended by state intervention or a legal liquidation. It can thus continue to exist almost in perpetuity. Yet, according to David Whyte, 'the contemporary form of the profit-making corporation is probably as close as we could get to a model organisation that is capable of destroying the world.' To a very large extent, capitalist corporations, in their endless pursuit of profit and capital accumulation, are prepared to destroy the sustainability of Earth's natural environments. Or, in effect, to commit ecocide, via: 'climate change, the ravaging of ecosystems, the eradication of species and the pollution of air, land and water.' Thus, confronting and defeating the capitalist corporation is crucial if we are to halt and then reverse the climate and ecological crises. This is why Whyte concludes that 'it is impossible to avert ecocide as long as corporations remain in control of the industrial processes that are wrecking our world.' However, Whyte is by no means alone in coming to such conclusions.[14]

Another who makes similar warnings about the nature of capitalist corporations is Grace Blakeley, who correctly points out that 'the point of a corporation is to maximise profits', and that 'corporations are a form of despotic private government' that is very difficult to constrain. This is especially true in relation to 'corporate sovereignty', where capitalist corporations are granted powers by governments which normally would only be exercised by the state alone. History has plenty of examples, stretching from the formation of the East India Company in the sixteenth century, right up to the various US corporations that were given power to run parts of Iraq after the US-led invasion in 2003. As Blakeley explains, corporations are, in effect, political entities, and were created as a legal concept 'to formalise certain relations of production, minimising bosses' exposure to risk, while maximising their ability to exploit and control their workers.' Essentially – and most importantly – the capitalist corporation was created as a '"structure of irresponsibility", designed to ensure "corporate impunity."' Such an entity is thus able to pursue its main purpose – amassing 'ever more wealth and power at the expense of people and planet' – with very little, or even no, accountability to society. The capitalist corporation thus has both economic *and* political power – power it uses to impose, and maintain, a very specific kind of production: 'one premised upon the domination of one class by another.' When such power is increasingly concentrated in just a few large multinational corporations, via mergers and take-overs, those corporations can then 'exert sovereignty in a similar way to states.' Thus, it is extremely difficult to control, or even limit, large capitalist corporations.[15]

Consequently, the capitalist system – the increasingly mis-termed 'free market' – via what is often chaotic competition between capitalist firms or countries, forces corporations to grow and become ever-more powerful, in order to maximize profitability and increase capital accumulation via increased production and consumption. Because of the power of such corporations today, they – and their capitalist system – become increasingly out of control: to the point where their continual push for economic 'growth' even destroys the basis of life on Earth. This was noted by Paul Sweezy in 1988, in 'Capitalism and the Environment', an article which he updated in 2004: 'The purpose of capitalist enterprise has always been to maximize profit, never to serve social ends.... Such is the inner nature, the essential drive of the economic system that has generated the present environmental crisis.' Such corporations have a long history: according to David Whyte, the world's first corporation – though not strictly a *capitalist* corporation – was a Swedish mining company called Stora Kopparberg, formed in 1288. Now known as Stora Enso, this corporation has become the world's largest paper producer. Although this corporation – in both its early form and its current one – has caused significant environmental pollution because of its operations, at no point has it had 'to face questions about its environmental record, let alone pay the costs of a clean-up or compensation for any damage it has caused.' As Whyte makes clear, 'this is the point of the corporation. It enables investors to walk away from the damage caused by their activities without ever having to face the consequences.' As he concludes, corporations like Stora show how 'no matter how much destruction corporations cause to us and the environment, they are designed to survive and thrive in perpetuity.'[16]

Soon, a fiction developed around such corporations in ways which asserted that the investors and shareholders were not the 'owners' of the corporation – the corporate 'person' was. A later example of such a corporation was the South Sea Company, formed in Britain in 1711, to extend control of the trafficking of enslaved people to Latin America. The investors were fully aware that they were investing their capital in a company which would ship almost 150,000 enslaved people a year from Africa to Latin America. Yet the legal existence of the corporation meant the investors could evade any 'moral accountability' and, instead, 'look the other way' whilst their capital grew from the exploitation of fellow human beings. It was to be corporations like this – such as the East India Company (formed in 1600), and the Royal African Company (formed in 1662) – that would soon extend a brutal and racist colonial expansion that eventually led to the formation of the British Empire. At the same time, similar corporations were formed in other European countries keen to establish their own colonies and empires. The ruthless ways in which such corporate colonization was carried out – displaying callous indifference to human suffering – also resulted in disastrous consequences for human ecosystems: 'Corporate colonisation was based on the enclosure of traditionally owned and managed land. This meant exerting absolute control over forests and farmlands, rivers and lakes.' Colonization, of course, was crucial to the birth of capitalism: 'global plunder, extraction and exploitation are at the root of the massive profit accumulations that made modern capitalism possible.'[17]

Under capitalism, from the mid-nineteenth century onwards, the corporation has developed into an entity which 'encourages indifference to human suffering and environmental degradation and ensures that social relationships between human beings are reduced to a financial relationship.' However, such behaviour by capitalist corporations continued well beyond the nineteenth century. Twentieth-century history gives plenty of examples of 'the collusion of major corporations with the most brutal of states... many of the largest household name corporations have collaborated and supported the most brutal and violent states and have participated in the most reviled acts of war and even genocide.' For instance, under the Nazi dictatorship, both the I. G. Farben and the Volkswagen corporations made extensive use of slave labour from concentration camps – in 1943, it's estimated that I. G. Farben were using over 150,000 such inmates, 'including around 30,000 from Auschwitz.' Nonetheless, both corporations were allowed to continue, in some form or another, after the Second World War, with 'many of the key managerial personnel and employees' keeping 'similar jobs doing similar things.' One of the main corporations of today's I. G. Farben group is Bayer – which, in the twenty-first century, 'has amassed the worst reputation for its impact on people and the environment.'[18]

Fossil fuel corporations – which play a crucial role in the economic life of most countries, and which have done so much to cause the current climate and ecological crises – also have a long and bad global history. In many ways, 'the corporation was formative in the development of a colonial capitalism that was always ecocidal.' As Mikaela Loach has shown, Shell was originally given a monopoly to 'extract from the oil-rich land' of Nigeria, after the British government had 'bought' the whole of present-day Nigeria from the Royal Niger Company. Shell then proceeded to completely pollute 'Ogoniland and the Niger Delta without any sense of guilt.' To this day, Shell still operates in the area, 'poisoning rivers so that they can no longer sustain marine life, filling the soil with carcinogens and toxins and destroying the homes and livelihoods of the local communities.' Whilst in Latin America, ChevronTexaco have been accused of committing genocide in the Amazon, in order to carry out oil extraction in a region in Ecuador. Their operations there have led to 'irrevocable damage to food and water supplies, the introduction of new diseases, and ultimately the extinction of entire tribes.' The conclusion to be drawn from those, and similar, examples is that such corporations continue to use their legal protections to act 'in ways that harm the environment or even kill and maim human beings as long as the economics makes sense.'[19]

Over the decades since the Second World War, corporations of all kinds – fossil fuel, tobacco, asbestos, synthetic chemicals and car – have manufactured deadly products. More importantly, many of the top corporate executives knew what they were doing, but 'wilfully ignored and actively sought to bury the evidence of their killer trade. They were fully aware of the consequences of what they were doing, but did it anyway.' For instance, most of the world's pharmaceutical and agrochemical firms are owned by three large corporations: Bayer is one, the other two being DuPont and ChemChina. An example of how such corporations can fairly be called 'killer corporations' is provided by DuPont, which manufactured, among other things, Teflon. DuPont's scientists – like those working for the fossil fuel and tobacco corporations – knew from as early as the 1960s that the chemicals used in the product were toxic. However, DuPont 'continued to dump the toxic waste left over from the production process into local waterways.' One chemical, known as C8, is linked to six serious human diseases, including two types of cancer – but was knowingly dumped into the water supplies of Parkersburg, a town of over 70,000 people. But it took years for the scandal to become public knowledge, and even more for compensation to be paid to the surviving victims – later becoming the basis of the 2019 film *Dark Waters*. To this day, it remains the legal aspects of the typical capitalist corporation which 'enables capitalists to disregard moral limits or indeed any limits on their profiteering that we might expect to be consistent with a basic sense of humanity.' Such corporations have dominated modern capitalism for over one hundred years, and have now developed into what Whyte describes as 'the deadliest human invention – an invention that has accelerated the capacity for the destruction of the planet in ways its creators could never have imagined.' As he concludes: 'we will not survive if we continue to allow corporations to occupy a central role in the economy.'[20]

The 'Corporation State'

In 1847, Marx and Engels, in *The Communist Manifesto*, had described the state in early industrial capitalism thus: 'The executive of the modern State is but a committee for managing the common affairs of the whole bourgeoisie.' This was an understanding shared later by many Marxist writers, including Daniel Guérin who stated, in his book *Fascism and Big Business* (first published in 1936), that: 'the state has always been the instrument by which one social class rules over other social classes.' In essence, capitalism is 'defined by the class divisions between owners and workers; between those who own all the stuff needed to produce commodities, and those forced to work to produce those commodities.' Because of this, those who 'own all the stuff' can influence politics in general – and even the state itself. After the Second World War, capitalists – and neoliberal ideologues – became increasingly hostile to Keynesian economics, welfare capitalism and reasonably-strong labour movements. From the mid-1970s onwards, neoliberals began pushing for 'the election of politicians who crushed the powerful labour movement, privatised public companies and marketized the welfare state.'[21]

As was seen earlier, neoliberalism is an ideology that is about much more than just markets and 'free' enterprise – at least as equally dangerous as its economic aspects, are its *political* impacts. Under neoliberalism, as all mainstream parties buy into its ideas and become part of the 'Extreme Centre', and democracy is thus 'hollowed out', states prioritize the economic interests of the one per cent. In essence, 'political power is captured by economic power.' As a consequence, state protections for the least well-off are removed; while policies to address increasingly-gross inequalities – via a progressive taxation system that tries to redistribute wealth – disappear even from the programmes of centre-left parties. Thus, like a vampire, as neoliberalism conquers the state, 'it sucks the power out of people's votes' – acting like 'a political neutron bomb' that seeks to replace politics with a 'democracy' based on consumerism. Although the visible signs of democracy may remain – such as multi-party systems, elections and parliaments – in reality, true political power has passed 'to other forums, inaccessible to ordinary citizens.'[22]

Under the sway of neoliberalism, states cease funding viable public services in order to avoid taxing the super-rich; instead, they privatize as many of those public services as possible. But, most dangerous of all, such states refuse to limit the actions of the large multinational corporations which are destroying so much of the living planet. It is this approach which has allowed capitalist corporations to get so dangerously 'out of control.' Partly as a result of political conviction – but also via the outright corruption of politicians from across the political spectrum – real decision-making has effectively shifted away from parliaments. Increasingly, crucial decisions are now taken in unofficial meetings between ministers and corporate lobbyists; in 'private meetings at economic summits'; in international trade treaties that contain Investor-State Dispute Settlement (ISDS) agreements which allow corporations to sue – in secret courts staffed by corporate lawyers – any state which dares to attempt to restrict their profit-making 'opportunities'; and even via the giving of 'donations' and 'gifts' to ministers. In these ways, under neoliberal capitalism, corporations can ensure that voter 'democracy' is unable to place any curbs on their anti-social greed and their ecologically-destructive operations. As George Monbiot and Peter Hutchison have said: 'democracy is the problem that capital is always striving to solve. Neoliberalism is the means of solving it.'[23]

For several decades, the power of corporations has grown to the point whereby the main functions of the state have been reduced to just two: 'further empowering the rich and crushing dissent.' This is why the number of 'illiberal democracies' and authoritarian states has actually *risen* during the period of neoliberalism's dominance – an economic philosophy that, at least according to the public statements of its creators, was supposed to increase democracy and so prevent the emergence of tyranny. According to the 2021 *Freedom in the World* Report, 'fewer than a fifth of the world's people now live in fully free countries.' According to that report, democracy had experienced such a serious deterioration since 2006 that the proportion of countries deemed 'Not Free' was higher than it had ever been, and that 'countries with declines in political rights and civil liberties outnumbered those with gains by the largest margin recorded during the 15-year period.' Their 2024 report pointed out that, by 2023, 'global freedom' had been declining for eighteen consecutive years, and concluded that: 'The breadth and depth of the deterioration were extensive. Political rights and civil liberties were diminished in 52 countries, while only 21 countries made improvements.' As climate protesters in the UK have found out during the past few years, civil liberties – relating to peaceful protest and even the right to a fair trial – have been significantly restricted by British governments and courts.[24]

As was seen in Chapter 4, capitalism and democracy have only rarely been natural 'bedfellows' – something noted more recently by Grace Blakeley in a June 2023 article: 'capitalism and democracy were never really all that compatible to begin with.' However, since the triumph of neoliberal capitalism from the 1980s onwards – and especially since the 2008 banking crash caused by the greed of large financial institutions – capitalism has reverted to its earlier attacks on democracy. Only this time, it has been able to 'capture' the bulk of state machinery in an increasing number of countries, in order to preserve and extend its global wealth and power. This is why, despite it becoming abundantly clear that serious measures are needed to reduce increasing economic inequalities and to tackle the climate and ecological emergencies, most states have failed for decades to do what is necessary. This is explained by going back to the understanding of the state as essentially the 'executive committee' of the capitalist class as a whole, which Marx and Engels put forward in the mid-nineteenth century. Historically, state institutions have never been 'independent entities that decide which action to take based on an objective assessment of the evidence.'[25]

This understanding is as true of the state bodies which ran the Roman Empire, and of the state institutions of the Soviet Union, Fascist Italy and Nazi Germany. Then and now, decisions result from political struggles within those institutions, with outcomes decided by the 'balance of power' within society as a whole. Not surprisingly, those with more wealth – and, thus, usually more power – tend to get their way more often than weaker groups. For example, despite climate experts warning, in very clear terms, that significant measures are needed to prevent climate and ecological breakdown, such measures have so far not been forthcoming. Even with a range of political leaders claiming they 'get it' as regards the climate crisis, what we see instead is that, globally, fossil fuel corporations receive a staggeringly-unbelievable $5.9 trillion in state subsidies. These are the same fossil fuel companies which, in March 2024, publicly derided efforts to move away from oil and gas at the industry's annual Cera Week conference in Houston (Texas) – 'despite widespread acknowledgement within the industry, as well as scientists and governments, of the need to radically reduce planet-heating emissions to avoid the worst effects of the climate crisis.' As Blakeley has argued: 'It is not a coincidence that an economy dominated by fossil fuel companies is also one in which action against climate breakdown has been slow and patchy: big businesses use their power *within* the state to make sure that the system benefits them.'[26]

The crucial aspect of the last quotation is the word 'within' – importantly, the power that corporations have over and within state institutions is not restricted to the fossil fuel giants which are continuing to use their power to prevent effective action on the climate crisis. Such power resides more generally with all the major industrial and financial corporations. All of the largest corporations are able to use their power to 'persuade' governments, to a greater or lesser extent, to allow them free reign in subjecting the citizens of the world to 'their ruthless free market competition'

– and to give them massive bailouts whenever their investments fail. The same holds true as regards weakening trade unions or worsening working conditions and safety protections: 'Businesses choosing profits over safety and governments bailing out bosses while letting people starve do not represent a perversion of capitalism. These things *are* capitalism.' Another, more general, example of how large corporations use their dominance of states is provided by the whole question of economic 'growth' – something essential for capitalism if it is to continue growing, and thus accumulating yet more capital. As Kate Raworth demonstrated so convincingly in her 2017 ground-breaking book *Doughnut Economics*, politicians of all mainstream parties – Ali's 'Extreme Centre' – have for many decades been committed to giving those corporations ever more 'growth'. Sometimes it's called 'sustained growth'; sometimes 'balanced growth' or 'long-term, lasting growth'; or even 'smart, sustainable, inclusive, resilient growth' – but, whatever adjectives are used, it's always that 'ubiquitous noun': growth.[27]

Such growth is usually measured in terms of Gross Domestic Product (GDP) — constant growth of which, over the past fifty years or so has increasingly come to be seen as a desirable political goal or even a political 'necessity.' Whilst there have been significant improvements in many aspects of human well-being since the end of the Second World War, the overall picture is 'patchy' at best. More importantly, it has become much more so since the fall-out from the 2008 financial crash, with a massive *increase* in social and economic inequalities. By 2015, 'the world's richest 1%' already owned 'more than all the other 99% put together.' However, continual 'growth' on a finite planet is, to say the least, highly problematic as regards both the climate and the natural world. If, as some hope and predict, global economic output grows by three per cent a year over the next twenty-five years, this would result in 'doubling the global economy in size by 2037 and almost trebling it by 2050.' This is what prompted Raworth to shift her focus from GDP to 'humanity's long-term goals' – and to come up with 'The Doughnut': 'a social foundation of well-being that no one should fall below, and an ecological ceiling of planetary pressure that we should not go beyond.' This concern with the latter aspect of continuous growth – the impact on the planet as a whole – was something which had been raised by the authors of the

1972 *Limits to Growth* report. This examined five global trends, three of which were 'accelerating industrialization, depletion of nonrenewable resources, and a deteriorating environment', all of which the report saw as being 'interconnected in many ways.' In 1999, in a lecture on Sustainable Systems, delivered at the University of Michigan, Donella Meadows (one of the 1972 report's authors) called 'growth' per se as 'one of the stupidest purposes ever invented by any culture', and asked the following pertinent questions: 'growth of what, and why, and for whom, and who pays the cost, … and what's the cost to the planet, and how much is enough?' As Raworth makes abundantly clear, the main task facing us in the twenty-first century now is 'to create economies that promote human prosperity in a flourishing web of life.'[28]

Winning a better future
What is currently preventing that vital goal from being achieved is the power of capitalist corporations – not just in the economy, but also over state institutions. This is why state power is so often and so consistently used to further the interests of large capitalist corporations, and of the capitalist system itself – even though *not* curbing their operations is increasingly endangering the conditions of life on Earth. However, this is part of the whole problem with capitalism – with its promoters, its enablers and its defenders. As Paul Sweezy has said: 'by far the largest part of the problem has its origin in the functioning of the world's [capitalist] economy as it has developed in the last three or four centuries.' Another to arrive at similar conclusions is James Gustave Speth, a leading 'insider' environmentalist who once acted as adviser to US presidents Jimmy Carter and Bill Clinton. In 1991, he had noted that while population growth had increased three-fold since 1900, 'economic output had increased twenty-fold.' Then, in his 2008 book, *Bridge at the Edge of the World*, he warned about the real dangers inherent in 'exponential economic growth, which is the driving element of capitalism', pointing out that 'the system had not delivered' the environmental reforms needed, and that 'capitalism as we know it today is incapable of sustaining its environment.' Since the end of the Second World War, 'growth' – as measured in terms of global GDP – has massively increased, 'at breathtaking speed.' Although this 'growth' has certainly been facilitated by the fossil fuel corporations, it's important to note that what has propelled such accelerated 'growth' has been the internal dy-

namic and logic of capitalism as a whole. As Murray Bookchin pointed out 'Capitalism can no more be "persuaded" to limit growth than a human being can be "persuaded" to stop breathing.' In a nutshell: the problem is capitalism itself, and how it has *always* functioned – not just the way it has developed in the past 50 years or so.[29]

And now, at last, for some 'good news' – the remaining chapters will explore how a different and better future from the one being created by capitalism – in particular, a fairer and more sustainable world – *is* both possible *and* within our grasp.

IN GREATER HARMONY WITH NATURE

For almost all of human existence, almost all of us were self-provisioning. Together with our neighbors, we lived and worked on the land, obtained and prepared our own food, and made our own homes, tools and clothing.[1]
Ian Angus

This book began with the Iroquois quote which stated that all important decisions should be taken with seven generations in mind – something that, as we have seen, capitalism tends not to do. The ecological irrationality of capitalism – or, in other words, its inability to live in true harmony with nature – is expressed by another Native American saying: 'Only after the last tree has been cut down. Only after the last fish been caught. Only after the last river has been poisoned. Only then will you realize that money cannot be eaten.' The essence of this view – attributed variously to either the Cree Nation or the Osage Nation – was more recently encapsulated by filmmaker, singer and activist Alanis Obomsawin, an Abenaki Native American from Canada. In a book published in 1972, after noting that Canada operated on a 'depletion economy which leaves destruction in its wake', she went on to say: 'Your people are driven by a terrible sense of deficiency. When the last tree is cut, the last fish is caught, and the last river is polluted; when to breathe the air is sickening, you will realize, too late, that wealth is not in bank accounts and that you can't eat money.'[2]

Remembering the past is important: not only can it pave the way to making important changes in the present – it can also help us see how to build a better future. It's now time to move on from 'pessimism of the intellect' – no matter how justified that pessimism is – towards building a vital 'optimism of the intellect' and, from that, the vitally-important 'optimism of the will'. One reason for having such justifiable optimism is because the *main* reason for all the crises we have examined in the preceding chapters is *not* because of so-called 'human nature' – even though that is what capitalists and their supporters have been trying to convince us, from the very beginning, is the case. *They* might be selfish, greedy and unmoved by evidence of human suffering, but most of us are not – witness only how, even during hard economic times, millions of ordinary people donate to emergencies, to help people even worse off than themselves.

Before capitalism

'To begin with' – to borrow three of the first six words from Dickens's *A Christmas Carol* – it must not be assumed that all pre-capitalist societies always lived in perfect harmony with nature. Just as nature itself has never existed in a permanent state of harmony and balance – with some of Earth's abrupt changes, such as volcanic activity, having pushed the environment to a state which presented serious challenges to existing life – so too did some pre-capitalist societies push their environments to the point of collapse.

One example often quoted is what is supposed to have happened on Rapa Nui (Easter Island), off the coast of Chile, about one century before the first arrival of Europeans in 1722. According to a long-established account, the Indigenous dwellers on the island, the Rapanui, over-exploited their natural environment, leading to hunger, warfare and, ultimately, to the collapse of their social and political systems. This was essentially the account put forward by Jared Diamond in his book, *Collapse: How Societies Choose to Fail or Succeed* (2005) – which dealt with several other cases of historical 'collapse', which actually were often the result of environmental factors. Some of those environmental issues were arguably self-inflicted – often via over-consumption of resources and consequent increases in waste-production and pollution of local habitats. In fact, as early as 1995, in an article entitled 'Easter Island's End', Diamond had said this of Rapa Nui: 'In just a few centuries, the people of Easter Island wiped out their forest, drove their plants and animals to extinction, and saw their complex society spiral into chaos and cannibalism. ... Eventually Easter's growing population was cutting the forest more rapidly than the forest was regenerating.' Towards the end of that article, Diamond asks a very pertinent question – which applies equally, if not more so, to the climate and ecological crises we're facing today: 'Why didn't they look around, realize what they were doing, and stop before it was too late?'[3]

However, more recently, the anthropologist Terry Hunt and the archaeologist Carl Lipo began arguing that, in fact, that there was no 'self-inflicted' ecological decline, and that, on the contrary, the Rapanui continued to live co-operatively in an environmentally-sustainable way. These two, along with Robert DiNapoli and Timothy Rieth, argued in an April 2020 paper that the evidence actually points to Rapanui society having continued to thrive, and even remain 'resilient…despite the impacts of European arrival.' According to Tom Garlinghouse, the evidence presented by these researchers suggests that, in fact, 'it was the newcomers from Europe who contributed to the island's societal collapse in the years to come.' Thus, rather than the Rapanui having brought about their decline by over-exploitation of their environment, it seems 'that the people of the island may have been the victims of European exploration and exploitation.' That, of course, is what happened in almost every case where European – and especially capitalist – countries first came into contact with Indigenous cultures in the Americas, Asia and Africa.[4]

Another ancient society that collapsed is the Mayan civilization. Diamond put this down to a combination of prolonged droughts because of climate change, and the destruction of too much of the surrounding forests. An article by Joseph Stromberg, in 2012, examined two studies which assessed those explanations. One study discovered evidence of 'severe reductions in rainfall… coupled with an [sic] rapid rate of deforestation, as the Mayans burned and chopped down more and more forest to clear land for agriculture.' A second study linked the decline in rainfall – sometimes failing to fall at all – to deforestation. According to that study, what was already a severe drought was exacerbated by the cutting down of trees – possibly with up to 60 per cent of the decline in rainfall being down to deforestation, which also contributed to soil erosion and depletion. Strangely, there's evidence to suggest that the Maya had 'developed a sophisticated understanding of their environment, [had] built and sustained intensive production and water systems and [had] withstood at least two long-term episodes of aridity.' Yet, despite understanding their environment and how to survive in it, they still 'continued deforesting at a rapid pace, until the local environment was unable to sustain their society.'[5]

However, it's important not to get too carried away with well-known dramatic instances of the collapse of pre-capitalist societies. For instance, both the Rapanui and the Maya continue to exist. As Professor Lisa Lucero points out: 'It was the Maya political system that collapsed, not [their] society. The over 7 million Maya living to-day in Central America and beyond attest to this fact.' This is an im-portant point, and one that's underlined by Professor Marilyn Mas-son who, in 2021, argued that 'Collapse is not a term that should be universally applied to "the" Maya, who should not be referred to as a single term either.' As she pointed out, the Maya region was large, with many different polities and environments – along with the Maya speaking a multiplicity of languages.[6]

This point about the Maya relates to an important aspect of pre-capitalist societies – whether hunter-gatherer, subsistence farming, or those with more developed agricultural systems. Whilst some did end up going beyond the ability of the local environments to sustain their civilizations at whatever level they'd reached, what often happened was that the people then moved to new areas – and applied the lessons they had learned from earlier mistakes. This was possible because the impact such societies had was restrict-ed to relatively-small local areas – unlike today's reach of global capitalism.

Stinting on the commons

Moving nearer to our times, the emergence of agricultural capital-ism – based on the private and exclusive ownership of land – was not a 'natural' process. This was something noted by Karl Marx in the first volume of *Capital*: 'One thing, however, is clear: nature does not produce on the one hand owners of money or commodi-ties, and on the other hand men possessing nothing but their own labour-power. This relation has no basis in natural history, nor does it have a social basis common to all periods of human history. It is clearly the results of a past historical development, the product of many economic revolutions, of the extinction of a whole series of older formations of social production.' As was seen in Chapter 6, the emergence of agricultural capitalism was achieved by a long process of illegal and legal robbery, often by violent means. This was just one – though extended – example of how external author-ities have been able to destroy communal systems, whether or not they were successful.[7]

So, what were some of those 'older formations of social pro-duction'; how successful were they; and how sustainable were they in relation to the natural world? As far as England is concerned, it would appear that agricultural systems based on commons-based farming were first brought over by Anglo-Saxon settlers, who arrived at approximately the same time that the Roman occupation was ending around 450 CE. There were different types of land-holding but, whatever forms they took, they soon became widespread, and even continued in various forms under the later feudal system.

Although, under feudalism, the land belonged to monarchs and powerful landlords, those who actually worked the land – most of whom were free peasants, with the rest being unfree serfs – had certain traditional and legal common rights to use some of the manor's resources. These customary rights were outside the control of the feudal lords, and were managed and allocated by those who worked on the land. Most villages seem to have used open-field systems, in which scattered strips of land were allocated – and periodically re-allocated – by the village community, so that each family would have its share of good and not-so-good soil. However, some areas in England and Scotland divided up communal land into more compact plots – a system that was known as *runrig* – but, these too were periodically re-distributed amongst community members.

While it's important to note aspects of social equality and democratic decision-making, it's also important to examine the environmental impact of such commons-based farming. As explained by Kate Raworth, 'natural commons have traditionally emerged in communities seeking to steward Earth's "common pool" resources, such as grazing land, fisheries, watersheds and forests.' Various studies have shown that such agricultural systems, based on various collective forms of ownership, were very much in tune with nature when it came to land-use and the importance of nutrients. Jeanette Neeson, for instance, discovered that common-field villagers in England – who met two or three times a year to decide issues – were very mindful of sustainability issues: 'Careful control allowed livestock numbers to grow, and, with them, the production of manure…Field orders make it very clear that common-field villagers tried both to maintain the value of common use of pasture and also to feed the land.'[8]

As Angus has shown, communal meetings held to discuss and decide future actions, and to deal with any problems, often concerned themselves with the question of 'stints' – these were 'limits on the number of animals allowed on the pasture, waste, and other common land.' Such measures were both democratic and ecological – by placing restrictions on the number of animals allowed on the commons, the villagers ensured that they remained healthy enough to accommodate the numbers each tenant was entitled to keep on it. Interestingly – in view of the later development of capitalist agriculture – such measures gave protection, especially to the poorer commoners, against the commercial activities of graziers and butchers who were more interested in making quick profits. Furthermore, the evidence suggests that such controls were successful – and so successful that, as Derek Wall points out, 'the use of stints continues for common land in England and Wales in the twenty-first century.' One example of such a modern-day commons is North Meadow in Cricklade, Wiltshire, which 'has worked well for over a thousand years, providing agricultural land while maintaining rich wildlife.'[9]

The whole question of keeping the soil healthy and productive was something common-field villagers were considering long before the so-called 'Agricultural Revolution' of the mid-eighteenth century. For instance, commoners controlled where sheep grazed, moving them by rotation from field to field, to ensure that all fields were sufficiently manured. Awareness of the crucial role of restoring nutrients to the soil also led to organized plantings of fodder crops such as turnips and clover in fallow fields, to make sure an optimum number of animals could be fed, along with obtaining increased amounts of manure. There is plenty of historical evidence to prove that 'long before scientists discovered nitrogen and nitrogen-fixing, these farmers knew that clover enriched the soil.'[10]

Interestingly, given the growing threat from pandemics – in part because of how industrial capitalism's intensive animal 'farming' increasingly concentrates animals in huge feeding 'facilities' – commoners were also aware of the dangers of spreading disease amongst livestock. For instance, at least from the early eighteenth century onwards, villagers drew up rules to isolate sick animals from the rest of the herd. In addition, they were also aware of the need to prevent farm animals from outside mixing with their own livestock. As well as present-day problems – such as the often-rapid spread of foot and mouth disease, swine flu and bird flu among farmed animals – a television documentary made by Queen-guitarist Brian May, and broadcast in August 2024, convincingly established that bovine TB is almost certainly spread, *not* by badgers, but by outside bulls being brought into herds for breeding purposes. That film also pointed out that bovine TB, to a large extent, was being spread *within* herds by keeping cows standing too long in their excrement. Centuries before, common-field farmers were concerned to stop livestock from fouling ponds. Furthermore – and again in advance of the eighteenth-century 'Agricultural Revolution' – these common-ers even developed 'strict controls on admitting bulls and rams to enter the commons for breeding', and showed an awareness of the need to prevent 'inferior' animals from breeding with their stock.[11]

It was in large part because commons-based agriculture was organized and managed by men and women who lived on the land that it tended to be ecologically-sustainable. The overwhelming bulk of the evidence suggests that common lands were well-man-aged, productive and sustainably farmed, and were passed on to the next generations in a good state. That is why this system survived as a way of community farming and living for so many centuries – and the fact that those common lands were so important for those who farmed them and lived off them explains why the resistance to capitalism's robbery of the commons was so determined and long-lasting. Although there were clear social divisions within such communities, the close and relatively-harmonious relationship with nature that they practized points, in a very strong way, to how such a way of living can be restored – both for people and for the rest of the natural world. It is no coincidence that Marx and Engels first understood how capitalism created a fatal metabolic rift between people and nature as a result of their study of the impacts of cap-italist agriculture. Although such commons-based agriculture was rarely conducted by the kind of egalitarian societies of 'associated producers' as envisioned by Marx, history shows that these com-moners nonetheless did 'govern the human metabolism with nature in a rational way.' In other words, in sustainable ways that stand in stark contrast to capitalism's ecologically-destructive methods.[12]

Wot 'Tragedy'?

However, not everyone has had such positive views of com-mon-based agricultural societies. One of the most influential – and determined – critics of such societies was Garrett Hardin. In 1968, he published the article 'The Tragedy of the Commons', which was largely concerned with his arguments in favour of strict controls on world population. While this was his main concern, the article also presented arguments against the viability of any system of agriculture based on the commons. According to him, as soon as social stability is achieved, what he argued was 'the inherent logic of the commons [would] remorselessly generate[s] tragedy.' This tragedy, according to Hardin, was the 'tragedy of freedom in a commons', which would see 'each herdsman [try] to maximize his gain.' According to Hardin, this was because each herdsman was what he called a 'rational being' whose main concern would 'always' be to increase benefits to himself by adding to the size of his herd – at the expense of overgrazing the commons. Even worse – because, according to Hardin, all the other herdsmen sharing the commons were also 'rational' – they would all do the same. Thus: 'ruin is the destination toward which all men rush, each pursuing his own best interest in a society that believes in the freedom of the commons. Freedom in the commons brings ruin to all.' Hardin applied this same 'logic' to pollution, as the 'rational' man 'finds that his share of the cost of the wastes he discharges into the commons is less than the cost of purifying his wastes before releasing them.' As all the other 'rational' commoners think the same, the commons end up polluted. However, his argument was based on the erroneous assumption that there was 'open access' to the commons – whereas the reality is that most commons were (and still are) in fact managed and governed in ways to ensure success by avoiding depletion of natural resources.[13]

Those who, on reading Hardin's arguments above, have been thinking that his view of the typical 'rational man' is very similar to pro-capitalist arguments that people are, by nature, greedy and selfish and thus think only of what benefits themselves, should award themselves a treat! Because, in that same article, Hardin goes on to say that: 'The tragedy of the commons as a food basket is averted by private property', which he believes 'deters us from exhausting the positive resources of the earth.' As Raworth correctly observed, Hardin's arguments fitted 'neatly into the neoliberal script.' Those who have some understanding of how capitalism has always operated, might thus find Hardin's conclusion somewhat surprising. Then, for most of the rest of the article, Hardin goes on to argue against 'the freedom to breed', in order to forcibly restrict population growth and to ensure the reproduction of those who are 'biologically more fit to be the custodians of property and power.' Such arguments seem to have been a natural progression in Hardin's thought in the years before 1968. Prior to that date, Hardin – who had 'no training in or particular knowledge of social or agricultural history' – was a professor at the University of California, and was best known for his biology textbook which had argued for 'control of breeding' for 'genetically defective' people. He continued to put forward such neo-Malthusian and eugenics-based ideas after publication of his 1968 article.[14]

More significantly as regards the commons, since 1968, Hardin's arguments have been influential in discussions about natural resource issues – and have 'been used time and time again to justify stealing Indigenous people's lands, privatizing health care and other social services, giving corporations 'tradeable permits' to pollute the air and water, and much more.' As Angus points out, 'the very fact that commons-based agriculture lasted for centuries disproves Hardin's assumptions.' This was also the view expressed by Susan Cox in her 1985 article, 'No Tragedy of the Commons', which argued that, on the contrary, far from there being a tragedy of the commons, there was in fact 'a triumph: … for hundreds of years – and perhaps thousands, although written records do not exist to prove the longer era – was managed successfully by communities.'[15]

Such systems of commons-based agriculture did not end because they were inefficient or unproductive – they ended only when those few with greater wealth and power grabbed the commons from the many. In fact, a recent paper on the global impact of capitalism, 'Capitalism and extreme poverty', convincingly demonstrates that, throughout history, such commons-based agricultural systems were 'innately capable of producing enough to meet their own basic needs (i.e., for food, clothing, and shelter), with their own labour and with the resources available to them in their environment or through exchange.' This paper, by Sullivan and Hickel, argue that there's considerable evidence to show it's unlikely that the vast bulk of people 'lived in extreme poverty prior to the rise of capitalism… Rather than being the natural condition of humanity, extreme poverty is a symptom of social dislocation and displacement.' It's natural disasters (such as droughts), or severe social dislocations (such as war, invasion or 'institutionalized dispossession' as under colonialism), which have usually tipped such communities into food shortages, famine or extreme poverty. Barring such occurrences, people living in commons-based agricultural societies have generally succeeded in adequately supporting themselves.

However, poverty usually arises when 'people are cut off from land and commons, or [when] their labour, resources and productive capacities are appropriated by a ruling class or an external imperial power.' While it's only fair to point out that significant poverty existed in pre-capitalist societies in which an upper class dominated and exploited the majority, it is also clear that this was not usually the case in societies based on systems of commons-agriculture. Rather, the inescapable conclusion is that it was 'the expansion of the capitalist world-system [which] caused a dramatic and prolonged process of impoverishment on a scale unparalleled in recorded history.'[16]

Commons: a 'natural economy'

Hardin's arguments about the alleged 'tragedy of the commons'were effectively rebutted by Elinor Ostrom – a political scientist who, in the decade following Hardin's paper, began researching 'real-life examples' of commons-based communities. In 1990, her influential book, *Governing the Commons*, was published. Her work on how commons-based societies – in Switzerland, Japan, Spain, the Philippines, India and the US – created and maintained 'well-managed natural commons' eventually won her an economics Nobel-Memorial prize in 2009, for 'demonstrating how local property can be successfully managed by a local commons without any regulation by central authorities or privatization.' She argued that commons – 'including common land, forests or fisheries that were owned collectively' – could be sustainably used and maintained. As Derek Wall argues, one of her most important findings – especially for the future – is that she demonstrated convincingly that 'democratic control, rather than top-down management or simple privatisation, works to conserve nature.' For instance, Ostrom and fellow researchers showed how farmer-built and farmer-run irrigation schemes for watering rice fields in Nepal were more successful in ensuring all farmers received sufficient water for their fields than those built and operated by the state. Even though the farmer-built schemes were less sophisticated than those built by the state, 'they were kept in better repair, produced more rice, and distributed the available water more fairly among all their members.' As Raworth explains, their self-organizing worked because the farmers involved 'developed their own rules for water use, met regularly in meetings and in the fields, set up a monitoring system, and sanctioned those who broke the rules.' Although she wasn't an anti-capitalist revolutionary, Ostrom's findings have effectively undermined arguments in favour of private ownership when it comes to protecting the natural world. She also found that this was because these successful commons were 'governed by clearly defined communities with collectively agreed rules and punitive sanctions for those who broke them.' This has confirmed the findings of historical studies which showed that, prior to the robbery of the commons, such agricultural communities were fully aware of the need to manage the commons in sustainable ways. Significantly, as regards building a better and fairer world, she discovered that such commons can become successful to the point of actually 'outperforming both state and market in sustainably stewarding and equitably harvesting Earth's resources.'[17]

The whole question of the environmental sustainability of the commons has implications that extend globally beyond the viability of local commons-based communities. This was something recognized by Ostrom who had spent so much of her time studying local examples of successful commons. In June 2012 – the same month in which she died – her article 'Green from the Grassroots' made 'a passionate call for action on global ecological problems.' In that article – published ahead of the UN's Rio+20 summit – she made it absolutely clear that action, on many fronts, was desperately needed over the coming decade, 'before the economic cost of current viable solutions becomes too high. Without action, we risk catastrophic and perhaps irreversible changes to our life-support system.' However, she warned against trying to draw up and implement a *single* plan for solving those problems. Instead, what she urged was an 'evolutionary policymaking' that was also 'green policymaking.' This would involve, amongst other things, building 'sustainable cities', which not only focused on being pollution-free, but which would also 'analyze flows of resources – energy, food, water, and people – into and out of their cities.' In addition, she argued – consistent with her work on the commons – that everyone, 'countries, states, cities, organizations, companies, and people everywhere', must be involved in setting goals for sustainability if action is to be successful. But she also noted – twelve years ago – that 'time is the natural resource in shortest supply.'[18]

Ostrom in effect 'advocated the seven-generation rule' and argued that, because 'long-term sustainability was essential', decisions should always be based on thinking of 'generations far into the future.' As Wall points out, she also correctly concluded that 'successful commons are nested in wider systems' – and those 'wider systems' include global elements: climate patterns and capitalist corporations, to name but two. Such global issues make modern-day commons – for example, Cricklade's North Meadow – vulnerable. Which is where Ostrom's approach falls a little short, because she – and her fellow-worker and husband Vincent – always maintained that capitalism, or 'market economies', 'could be embedded in ecologically sustainable systems.' In part, this was because she believed that the activities of corporations that attempted to strip 'open' resources – such as the oceans – in the pursuit of short-term profits, could be controlled by 'better governance.' Sadly, the evidence suggests that the biggest corporations – and, arguably, the capitalist system itself – are beyond any form of 'control'.[19]

Essentially, a commons-based economy is what Marxists have sometimes called a 'natural economy', which capitalism has to destroy if it is to make profit from the land. As Rosa Luxemburg explained, 'forms of production based upon a natural economy are of no use to capital.' This is because in 'all social organisations where natural economy prevails' – usually in 'peasant communities with common ownership of the land' – economic developments are 'essentially in response to the internal demand… and therefore there is no demand, or very little, for foreign goods, and also, as a rule, no surplus production.' It was this attitude towards 'natural economies' which, in the second-half of the nineteenth century, led to Britain, and other developing capitalist nations – including the US – relentlessly waging aggressive wars against those commons-based societies and cultures which still existed. By then, such ways of living were mainly restricted to Indigenous peoples, who were conveniently termed 'primitive' or 'uncivilized' by 'modernizing' proponents of capitalism, to 'justify' their actions. The robbery of Indigenous lands which followed mirrored the earlier destruction of the commons in Europe, but was often more brutal. Unsurprisingly, these capitalist and colonial onslaughts were, as earlier in Europe, often resisted.[20]

Indigenous protectors
In fact, probably some of the best examples – both past and present – of how pre-capitalist civilizations have lived, successfully, in greater harmony with nature, are provided by Indigenous peoples.

As Wall has noted, Indigenous people, 'have often maintained commons for hundreds or even thousands of years without destroying these environments.' Now, present-day Indigenous people are being increasingly seen as 'Protectors' of what's left of the natural world. It is the past, and present, relationships of Native Americans with nature which forms part of Professor Nick Estes's 2019 book, *Our History is the Future* – and it is such Indigenous attitudes to the natural world which offer important ways forward, today, to re-establishing ways of living and thriving that are much more in harmony with nature. Indigenous peoples had – and still have – a relationship with the natural world which tends to see all the different parts of it as animate, rather than inanimate – and even as 'relatives' who should be respected, and treated well. Such a relationship which 'has been derided for centuries as "primitive superstition" is now increasingly seen as an important type of understanding and knowledge. The Lakota and Dakota Nations, for instance, lived by the philosophy of Mitakuye Oyasin, meaning "all my relations" or "we are all related."' This applied not just to human relatives, but also to all non-human or 'other-than-human life.' For the Oceti Sakowin (Sioux), the 'highest insult' was to be called 'Wasicu' – this translates as 'the fat taker', and applied to those who 'behaved selfishly, individualistically, with no accountability, as if they had no relatives.' Such people were usually punished by temporary shunning or alienation, or in worst cases, by banishment from the tribe. As was soon discovered by the Native Americans, 'Wasicu' precisely encapsulated the approach of European colonialists and capitalists.[21]

According to Estes, 'Racial capitalism was exported globally as imperialism, including to North America in the form of settler colonialism.' In the nineteenth century, this was ramped up, with US capitalism inflicting an 'apocalyptic death world' on ecosystems and Native American Nations. During the relentless wars waged by the US government and its European citizens against Native Americans, there were frequent attempts to agree treaties – described by one Indigenous leader as 'thief treaties'. During the discussions preceding the signing of such treaties, the words of Indigenous leaders were recorded – and provide a clear insight into how Native American Nations viewed, and lived with, the non-human natural world. Sadly, as we know, those treaties were often broken quite quickly, as more and more European settlers poured into the lands where Native Americans lived. That entire process was later summed up by Makhpíya Lúta (Red Cloud) of the Oglala Lakota Nation: 'They made us many promises, more than I can remember, but they never kept but one; they promised to take our land, and they took it.' During the relentless wars of robbery and genocide inflicted on Native Americans – who often also waged short wars against each other – their attitudes to the land and nature were revealed by several Indigenous leaders. In 1875, when yet another treaty was abrogated by a presidential proclamation, Heinmot Tooyalaket (Young Joseph) of the Niimíipuu (Nez Perce) people, declared that: 'The earth was created by the assistance of the sun, and it should be left as it was…The country was made without lines of demarcation, and it is no man's business to divide it.' A similar, non-capitalist, view of the natural world was expressed by Tashunka Witko (Crazy Horse) of the Oglala Lakota: 'One does not sell the earth upon which the people walk.'[22]

Indigenous people like the Lakotas never saw themselves as *owning* the forests or rivers, instead they believed it was they who belonged to the natural world. Thus – like the European commons with their self-imposed stints – while no one was barred from hunting, 'they did place restrictions' on hunting simply for profit. Hence it is not true, as claimed by some, that Indigenous people themselves were largely to blame for the decline of beavers, buffalo, and other animals – it was instead down to the industrial-scale killing used by European companies and individuals in order to reap quick profits. Once beavers had been all but exterminated by such methods, these groups moved on to buffalo – though they only wanted the hides and, later, the tongues. However, such prolific waste went further: 'white hunters not only left buffalo carcasses to rot on the plains, but also poisoned them with strychnine to kill off coyotes, wolves, and other scavengers – and sometimes starving native people.'[23]

On a separate occasion, Wanigi Ska (White Ghost) of the Yankton Lakota, spoke for many when he described what the white settlers were doing to the lands on which his people lived: 'You have driven away our game and our means of livelihood out of the country, until now we have nothing left.' This was actually part of a deliberate policy by US authorities to drive Native Americans into small reservations, by depriving them of game to hunt. However, that policy also illustrated a big divide over how to treat the natural world and the non-human animals which lived in it. On 9

November 1867, Bear Tooth, a chief of the Apsáalooke/Absaroka (Crow) Nation, condemned Europeans for their reckless destruction of wildlife and the natural environment: 'they have destroyed the growing wood and the green grass; they…have devastated the country and killed my animals, the elk, the deer, the antelope, my buffalo. They do not kill to eat them; they leave them to rot where they fall.' However, this deliberate 'scorched earth' policy continued. In 1865, it was estimated that there were approximately 15 million buffalo remaining from an original number of 25-30 million – by 1872, only seven million were left. According to the 1948 UN Convention on the Prevention and Punishment of the Crime of Genocide, the deliberate destruction of resources 'needed for physical survival, like clean water, food (e.g., buffalo), clothing, shelter or medical services' is seen as genocide. Then, from 1872-74, another 3,700,000 buffalo were killed, mainly by buffalo hunters and the US Army. During that two-year period, only 150,000 buffalo were killed by Native Americans as part of their usual subsistence-hunting. Not surprisingly, as noted by Dee Brown, this wanton destruction and waste meant that Native Americans 'wanted no part of a civilization that advanced by exterminating useful animals.' Sadly, what happened to the Native Americans – who, for centuries, had successfully established harmonious relations with the natural world and had thereby maintained sustainable commons – is what has happened to so many peoples and lands which were 'taken over by external authorities and destroyed.'[24]

Indigenous people have also tended to have markedly different attitudes to the poor. This was something reported by Graeber and Wengrow, in their 2021 ground-breaking book, *The Dawn of Everything*. When Kandiaronk, a Native American chief of the Huron-Wendat Nation, visited European cities like Montreal, New York and Paris in the late-seventeenth/early-eighteenth centuries, he was shocked by how wealthy Europeans could walk past poor people and beggars – who were clearly in 'abject poverty' – without sharing their wealth with them: 'Do you really imagine I could carry a purse full of coins and not immediately hand them over to people who are hungry?' As Graeber and Wengrow state, even Europeans who'd originally been taken captive by Native Americans in raids usually chose to remain in Native American communities, for a variety of reasons – including their 'reluctance ever to let anyone fall into a condition of poverty, hunger or destitution.' Later, Thathánka Íyotake (Sitting Bull) of the Lakota Nation – who played a big part in Custer's defeat in 1876 – joined Buffalo Bill's Wild West Show in 1885. However, when members of the crowds later gave him money for signed photographs, he 'gave most of the money away to the band of ragged, hungry boys who seemed to surround him wherever he went.' He once told Annie Oakley that he failed to understand how 'white men could be so unmindful of their own poor. "The white man knows how to make everything, but he does not know how to distribute it."' It would appear that Native Americans 'found life infinitely more pleasant' in societies where no one was 'in a position of abject poverty.' Such different values – on sharing, as well as on power and rules – perhaps helps explain that when Europeans were adopted, or even captured, by Indigenous groups, they rarely wanted to return to European 'civilization.' Some of those Europeans cited a variety of reasons for their decision to remain, such as 'the …freedom …, including sexual freedom, but also freedom from the expectation of constant toil in pursuit of land and wealth.' This is in stark contrast to Native Americans who'd become part of European communities (whether willingly or otherwise) – these almost invariably chose to return to their own people.[25]

Capitalism vs *Pachamama*

Yet, despite US governments and military inflicting such genocidal policies on Native Americans, the latter's traditional – and more sustainable – attitudes to the natural world have survived into the twenty-first century. Ecological economists – like Julia Steinberger and Kate Raworth – argue convincingly that what the world urgently needs is an economic and social system based on ecological principles. This is essential in order to sustain nature and its ecosystem services if we are to maintain a decent quality of life for us and for future generations. Such an approach is what remains at the heart of Indigenous cultures. However, as Robin Wall Kimmerer, and others, have pointed out, Indigenous understandings of the natural world and all the species that are part of it, and depend on it, continue to run completely counter to the daily operations of capitalism. Today, it's neoliberal capitalist 'philosophy', not an Indigenous philosophy, that rules. The former operates on the assumption 'that human consumption has no consequences.' Essentially, capitalism remains wedded to what is, in effect, a 'death cult' myth that it's possible to have infinite growth on a finite planet, 'as if somehow the universe has repealed the laws of thermodynamics on our behalf.' As has been stated earlier in this book, perpetual 'growth' is simply not compatible with nature. Yet, as Kimmerer has pointed out, this doesn't stop leading pro-capitalist economists from constantly stating, in different ways, the nonsensical 'argument' that 'there are no limits to the carrying capacity of the earth that are likely to bind at any time in the foreseeable future. The idea that we should put limits on growth because of some natural limit is a profound error.'[26]

On several occasions – as noted previously – words such as 'monster', 'vampire', 'vulture' and 'juggernaut' have been used to describe how capitalism operates in such a way as to preclude a harmonious and sustainable relationship with the natural world. To this day, the Native American Anishinaabe people have tales about a 'legendary monster' called the *'Windigo'*. As explained by Kimmerer, the *'Windigo'* is 'the villain of a tale told on freezing nights in the north woods.' Interestingly, this monster is in many ways a metaphor for how capitalism lacks harmony with the natural world. This monster is over ten feet tall, and hunts for its victims in winter: 'the hungry time…[and] Most telling of all, its heart is made of ice' In particular, according to the legend, the Windigo suffers 'a hunger that will never be sated. The more a Windigo eats, the more ravenous it becomes.' As a consequence, 'Consumed by consumption, its lays waste to humankind.' Significantly, this legend seems to have become widespread during the time of the capitalist fur trade, which drove so many species to near-extinction, and which consequently brought famine to many Native Americans communities. Creation stories and myths – including tales of monsters like the Windigo – often reflect aspects of people's deeply-held world-views, core-values and philosophies: of how they see their place in the wider world, and how they should relate to it. As Kimmerer says: 'Windigo is the name for that within us which cares more for its own survival than for anything else.' Thus, in terms of establishing harmony with the natural world, the Windigo can be seen as a destructive 'positive feedback loop', in that it shows how 'uncontrolled consumption' is undesirable and leads to negative consequences. Kimmerer sees the Windigo legend as a way to encourage 'negative feedback loops' that produce balance and thus help promote the maintenance of a stable system., by building 'resistance against the insidious germ of taking too much' from the natural world of which are just one part, and on which our survival depends.[27]

There are, of course, other 'Indigenous Protectors' in other continents; and, increasingly, the vital role they play in sustaining what remains of the natural world is being recognized by individuals and international bodies. Elinor Ostrom, for instance, was one to do so, arguing that 'indigenous peoples who protected forests that acted as a carbon sink should be supported.' In 2021, at COP26 in Glasgow, the important role played by today's Indigenous people, and their fundamental attitude to the natural world, was made abundantly clear by Narubia Werreria, an environmental activist of the Amazonian Karajá people: 'Indigenous peoples are the true guardians of the rainforest, its last line of defence…We are not the owners of the Earth, she owns *us*. We need a relationship not of domination, but of mutual affection. …We have to unite, and this alliance must spread all over the Earth.'While, in 2022, the Democratic Republic of the Congo passed a law – which came into force in early 2024 – that recognized the 'customary and land rights' of the Indigenous Pygmy peoples. In particular, the law recognized 'their crucial role' in protecting what now constitutes 30 per cent of the planet's oceans, lands and freshwaters. Increasingly, such groups of Indigenous people are being awarded significant grants – by organizations such as Green Grants and Nia Tera – to help them carry out their valuable work protecting the natural world. In 2023, Nia Tera gave out $24.3 million in grants to 128 organizations supporting over 270 Indigenous peoples around the world.[28]

In many ways, the attitudes of Indigenous peoples to the natural world have affinities with ecofeminism – or ecological feminism. Ecofeminists tend to see the devaluing, and the resulting degradation, of nature as arising in large part from the same dualistic Cartesian philosophy (itself linked to Plato's division of the world into 'spirit'/mind and 'matter'/nature) that also results in patriarchy and the oppression of women: the linked ideas that Nature is inferior to 'Reason', and that women are 'close to nature' and thus 'inferior' to men who are said to be 'rational'. Since the term 'ecofeminism' was first coined by Françoise d'Eaubonne, ecofeminists have argued for the creation of a society free from hierarchy – which no longer sees all 'Earthlings' as 'Others', and in which interactions are based on the recognition that all living beings are part of that common organism we know as Earth In 1993, Val Plumwood's book *Feminism and the Mastery of Nature* pointed out that how feminist utopias often describe 'a land where women live at peace with themselves and with the natural world.' Such a land would be without hierarchy – whether among humans, or between humans and animals. Instead, it would be a world 'where people care for one another and for nature, where the earth and the forest retain their mystery, power and wholeness, where the power of technology and of military and economic force does not rule the earth.'[29]

Mention should also be made of those who have joined with Indigenous peoples to protect their ways of life from capitalist destruction of their environments. One important campaigner was Hugo Blanco, a Peruvian Marxist, who died in 2023. During the 1960s, he helped lead a peasant uprising which was largely successful in achieving significant land rights – causing Che Guevara to say, in 1963, that Blanco was setting a 'good example' because he struggled 'as much as he could', despite strong repression. According to the historian Eric Hobsbawm, 'Hugo Blanco helped "the most important peasant movement of [the 1960s] in Peru, and probably in the whole of South America."' At the time, Blanco was a leading member of the Trotskyist Fourth International, and maintained comradely relations with the Fourth International until his death in 2023. According to leading UK ecosocialist Derek Wall, Blanco was 'a pioneering ecosocialist, promoting an ecological approach to revolutionary activism before many of us were conscious of this element.' Towards the end of his life, Blanco also supported the Zapatistas in Mexico, and other Indigenous struggles around the world – all in defence of Pachamama (Earth Mother). His commitment to environmental issues never wavered. In 2017, he said he was with 'the Amazonians who fight in defence of the rainforest, which are the lungs of the world. I am also against agro-industry.' According to him, 'while indigenous people might not use the term "ecosocialist", they have been fighting for ecosocialism for five hundred years.' Ultimately, he believed 'that the ecological crisis is an immediate threat to humanity and the focus of struggles must be to end the capitalist system that generates this threat.'[30]

However, as has been seen previously, it is often extremely dangerous to support Indigenous peoples in their struggles to protect their environments from capitalist destruction – and several have lost their lives in those struggles. Those who have paid the ultimate price include two campaigners in Brazil: Chico Mendes and Dorothy Stag. Mendes was a rubber-tapper who, in the 1960s, became an environmentalist and also embraced socialism (or, according to Alan Thornett: 'eco-Marxism'). In the mid-1970s, Mendes – along with Wilson Pinheiro – began organising mass resistance against the deforestation being carried out by huge logging companies. In 1980, Pinheiro was murdered by a death squad; but Mendes carried on, and formed the Forest Peoples Alliance. This brought together, for the first time, rubber-tappers and Indigenous people to defend the Amazon rainforest against deforestation. In 1988, Chico Mendes was himself shot dead by a gunman hired by the big landowners. However, his murder did not stop others from continuing to support Indigenous and peasant campaigns. Particularly important more recently was Dorothy Stag, who helped local peasant farmers who were being attacked by criminal gangs hired by big ranchers to grab their small plots of land. In June 2004, she gave evidence to a national Commission of Inquiry on Rural Violence, testifying against the big landowners who were, quite literally, getting away with violence and murder. Just eight months later, at the age of 73, she was murdered by a death squad. But, although many more of those who continue to struggle in defence of Indigenous groups, and against the destruction of the rainforest, have since been murdered, those eco-heroes – as is often the case – continue to inspire resistance.[31]

Calling time on capitalism's *Windigo*

For humans to survive and thrive, they must interact with the natural world in a much-more harmonious and sustainable way. Because, ultimately, we are part of – and thus dependent upon – Earth's ecosystems. Achieving that greater harmony means recognizing and accepting that we live in a finite world, which has limits and constraints as well as possibilities. In the past, some human societies have adopted interactions that pushed isolated local environments beyond their natural limits, thus leading to situations which threatened human survival – in those *local* areas. The big difference between such societies and today's capitalism is that the former were local, whereas capitalism's global reach now threatens global mass extinction.

However, as Rosa Luxemburg pointed out, such 'natural' economies and societies stand in direct opposition to capitalism, because such an economy 'confronts the requirements of capitalism at every turn with rigid barriers.' As a result, capitalism – wherever it exists – must always fight 'a battle of annihilation against every historical form of natural economy that it encounters.' Chief Arvol Looking Horse of the Lakota Nation – who played a leading role, from 2016 onwards, in resisting the Dakota Access Pipeline (DAPL) – voiced a traditional Indigenous view which, like Luxemburg's argument, runs directly counter to capitalism's priorities, but which has continued to this day: 'To us, as caretakers of the heart of Mother Earth, falls the special responsibility of turning back the powers of destruction.' Today, the 'colonized and racialized poor are still burdened with the most harmful effects of capitalism and climate change, and this is why they are at the forefront of resistance.' Now, according to Estes, 'Indigenous and non-Indigenous people will have to unite to turn back the forces destroying the earth – capitalism and colonialism' if, on a *global* scale, we are to protect what the Oceti Sakowin (Sioux) Nations call 'Unci Maka': 'Grandmother Earth.'[32]

In 1722, the Haudenosaunee Indigenous people formed the Iroquois Confederacy of five (later six) Nations in the north-east of North America. They, like many other Indigenous peoples, believed that as the natural world sustained them, they had a duty to be respectful of nature and its gifts, and, to care for and so help sustain the natural world on which they depended. Such groups still believe 'that, in return for the gifts of Mother Earth, human people have a responsibility for caring for the nonhuman people, for stewardship of the land.' The Onondaga Nation, for instance, believe they are 'one with the land and consider themselves stewards of it.' For them, the nation's political leaders have the duty to work 'for a healing of this land, to protect it, and to pass it on to future generations.' In the US, in late 1880s, there arose the 'Ghost Dance' phenomenon, led by Wovoka, a Paiute holy man who had had a vision of when the traditional harmony between humans and the rest of the natural world would be restored by the return of 'correct relations among human and nonhuman worlds.' Today, many see such aims as an 'part of a growing anticolonial theory and movement' that can help restore such a greater harmony between humans and the non-human parts of the natural world.[33]

To do that, we need to start believing in ourselves, as well as giving due respect to the natural world. For it is not humans, as such, who're responsible for the current ecological and biodiversity crisis – the fault lies with the destructive tendencies of the capitalist system, and the nature – and people-destroying greed of a very wealthy and powerful, but very small, minority of humans. As Julia Steinberger has argued: 'Quite simply, … [t]he state & trajectory of our current societies do not reflect the aspirations, potential or desires of the vast majority of our fellow humans.' Although capitalists, and those who defend their system, are at pains to depict all humans as being like themselves – 'isolated, selfish, cut-throat competitive' – this is *not* in fact what most of us are. In order to survive, humans have had to be co-operative, caring and sharing; and history has shown that humans have created societies, based on harmony with their environments, that have lasted for hundreds and even thousands of years – and that most of these only ended as a result of the onslaughts of capitalism.[34]

In addition, researchers like Graeber and Wengrow have shown that, even after the advent of agriculture, humans have been able to develop cultures that were both democratic and equitable: 'Agriculture, … did not mean the inception of private property, nor did it mark an irreversible step towards inequality.' Evidence also shows that even with the move to living in cities, a good number of the earliest ones established were 'organized on robustly egalitarian lines, with no need for authoritarian rulers, ambitious warrior-politicians, or even bossy administrators.' Thus, with world history presenting so many hopeful examples of human societies, it is way past the time for countering the negative myths of capitalism with the understanding – based on historical and contemporary *fact* – that humans are more than capable of creating societies that are capable of living in harmonious ways with the rest of the natural world, which are fundamentally democratic, and which are based on equity and taking care of each other. Because, after all, humans are 'projects of collective self-creation.' As Steinberger concludes: 'This is who we were, and who we can become again. It is time to encourage ourselves and our fellow humans to believe in our collective capacity to change things, work & care for each other's prosperity, and kick the neoliberal monsters devouring our societies out of our history.'[35]

A BETTER WORLD *IS* POSSIBLE

We need to remember that the work of our time is bigger than climate change. We need to be setting our sights higher and deeper. What we're really talking about, if we're honest with ourselves, is transforming everything about the way we live on this planet.[1]
Rebecca Tarbottom

As the above quotation – from Rebecca (Becky) Tarbotton of the Rainforest Action Network – recognizes, fundamental changes are needed to ensure that all humans – and non-humans – can thrive sustainably on planet Earth. Changes both to slow the destructive trajectory the world is currently on, and, eventually, to achieve such a great transformation – or 'System Change' – that we can reach a stage where those initial reforms cannot, like previous ones, be overturned later by wealthy and powerful minorities. Whilst what prospects there ever were, of a 'green' capitalism, are rapidly evaporating, it's nonetheless important to push for some immediate reforms. In part, this is because we desperately need to win time to mitigate the harms currently being inflicted on the natural world by capitalism. As Löwy has said: 'The struggle for ecosocial reforms can be the vehicle for dynamic change, a "transition" between minimal demands and the maximal program.'[2]

Basically, there are two kinds of reforms: those that improve situations but don't fundamentally challenge business-as-usual, and those which begin to change people's ideas *and* weaken the existing system. The latter are often called 'transitional reforms', as they try to achieve two kinds of goals: immediate improvements in current circumstances, along with preparing the way for fundamental and radical change. However, the former should by no means be dismissed out of hand simply because they don't bring about truly radical changes.

Whilst Starmer's version of Blair's 'New Labour' is far from ideal – whether on support for Israel's war crimes against Palestinians, the two-child benefits cap, the winter fuel payments to pensioners, or on migrants and refugees – to argue that the change of government in the UK, following the July 2024 election, has made no difference at all, is to ignore some small but significant gains. Had the Tories stayed in power, it's very unlikely they would have decided against issuing any more *new* oil and gas licences in the North Sea, or would have pulled their legal defence for a new coalmine in Cumbria. In addition, the new Labour government still seems to be committed to de-carbonizing the National Grid by 2030, and has also set up GB Energy to increase production of green and cheaper renewable energy. Then, in November 2024, it announced that it will ban the licensing of new coalmines in the UK.

We should celebrate any such gains, even small ones, as victories – and then use them to press for more. That reforms – even if they don't fundamentally challenge the system – are nonetheless important, has been recognized even by those who, like Paul Sweezy, are committed to replacing capitalism with an ecologically-sustainable and socially-just system. As he pointed out, conservation movements emerged in most capitalist countries during the twentieth century – and many managed to impose some limits on the more destructive depredations of uncontrolled capital. His conclusion on such reforms was unambiguous: 'It is hardly an exaggeration to say that without constraints of this kind arising within the system, capitalism by now would have destroyed both its environment and itself.'[3]

This chapter will offer some pointers to how we can get to a better and fairer world – by considering three ways that offer improvements, for the majority, in the present *and* which also help weaken the destructive capitalist system; and by suggesting three changes that could help in building an ecologically-sustainable and socially-just world.

Weakening the 'Demon Destroyer'

While we should stay clear of science-fiction 'unicorn' geo-engineering schemes – designed to allow capitalistm's destructive business-as-usual to continue – there are several key aspects we should push for and engage in: actions to block new fossil fuel or nature-destroying operations, policies that make fossil fuels more expensive to polluting corporations, and reforms that improve lives *and* weaken the capitalist system and so take us nearer to a better future.

Blockadia

One important type of action is 'blockadia' – a term coined by Naomi Klein in 2014, to describe those who, quite literally, put their bodies on the line to block fossil fuel projects, mining operations, the destruction of rainforests and mountain habitats by agri-businesses, or the pollution of rivers and seas. For several years, in a large number of countries, such activists have been arrested and imprisoned, or even killed, for resisting capitalism's life-destroying activities. In 2017, Global Witness published their 'Earth Defenders' global database, which showed that 'the number of environmental activists killed had gone up from 1 to 4 a week over just ten years', and that 2017 had already been '"the deadliest year on record" for environmental activists.' Between 2012 and 2023, more than 2,100 land and environmental defenders had been killed for defending their lands; and in October 2024, it was reported that an Indigenous Earth defender was now being killed every four days.[4]

Despite this, climate and ecological warriors have in fact had some notable successes around the world – for instance, 'blockadia' and anti-extractivism methods played a huge part, after more than ten years of popular resistance, in the cancellation of the Keystone XL pipeline by US president Joe Biden in 2021. In the UK, one successful example of 'blockadia' was resistance to the fracking operation Cuadrilla wanted to carry out at their Preston New Road site in Lancashire – which, if successful, would have been the first example of horizontal fracking in the UK. There, local campaigners, organizations such as Reclaim the Power, and the Anti-Fracking Nanas, joined together over several years to prevent this project. A variety of methods – including physical ones such as 'slow-walking' in front of trucks, truck-surfing, lock-ons, and sit-downs – meant that the company was unable to implement its plans. Another example of the success of blockadia methods was the role played by Coal Action Network in stopping further coalmining in the Pont Valley after 2020. Such actions – whether they stop or merely slow fossil fuel projects – are important. This was noted in 2017 by Professor Joan Martínez-Alier: 'The fight against climate change requires that a large percentage of the recoverable fossil fuels remain in the ground. These are the so-called *unburnable* fuels.' The importance of 'non-state actors' in blocking fossil fuel operations has also been recognized more recently by Malm and Carton, who point out that each success in blocking such operations was 'proof that projects could be thwarted, and that even if the enemy had *mostly* not ceased to be victorious, it was by no means invincible.' In addition, Extinction Rebellion, Just Stop Oil and Youth Demand activists have also put their bodies on the line – often suffering arrest and, increasingly, imprisonment – by engaging in peaceful direct actions designed to push the climate and ecological crises higher up the political agenda.[5]

However, many blockadia campaigns have also used other methods, including successfully using the courts. In the UK, this led – after a long campaign by the Weald Action Group – to the Supreme Court ruling, in June 2024, against fracking for oil at Horse Hill in Surrey. Following on from that victory – described as 'a heavy blow for fossil fuel lobbyists' – in September 2024, the High Court ruled against the Cumbrian coalmine in a case brought by members of South Lakes Action on Climate Change (SLACC) and Friends of the Earth. Such decisions help in the much-needed transition away from fossil fuels, and instead increase pressures for investment in clean, renewable energy. As one of the lead people bringing this case, Maggie Mason of SLACC said: 'the decision could "advance the global phase-out of fossil fuels."' Not surprisingly, such legal victories have led to further court cases against projects already approved. For instance, Greenpeace is amongst those fighting a court case against the development of Rosebank, 'the UK's largest untapped North Sea oilfield.' No doubt as a result, the new UK government recently decided to pull the legal defence of the Rosebank project, and also of the Jackdaw project. As Greenpeace stated, though it's not 'completely over yet… this is a massive step forward.'[6]

A combination of such blockadia methods – whether involving 'people marching, petitioning, litigating, divesting, sitting in, blockading and engaging in other direct actions' – continues to be hugely important in the struggle to avoid climate and ecological collapse. A study – published in 2020, of nearly 400 cases around the world between 1997 and 2019 – showed that 'over a quarter of projects encountering social resistance' had been blocked. That 'social resistance' led to the following results: 15 per cent of pipelines, 26 per cent of fracking projects, and 18 per cent of other oil and gas projects, were 'cancelled, suspended or had their investments withdrawn.' Overall, with coalmining projects included, the average success rate of blockadia methods for stopping fossil fuel projects was 25 per cent. As each year goes by, such 'social resistance' becomes increasingly crucial – because, if we are to avoid exceeding 1.5°C of global warming above pre-industrial levels, there needs to be both a rapid phaseout of fossil fuels, and a massive expansion of renewable energy.[7]

In 2021 – coming after the drop in CO_2 emissions caused by the temporary COVID-19 slowdown – the *Nature* magazine had surveyed IPCC authors and editors to find out how much they now expected the world to warm. Of the 92 who responded, 60 per cent thought it 'most likely' that by 2100, global warming above pre-industrial times would be 3°C or more; only 4 per cent 'believed the world would limit warming' to 1.5°C. In addition, 82 per cent of the respondents thought they would see 'catastrophic impacts of climate change' in their lifetimes. In May that year, the International Energy Agency (IEA) pointed out, that in view of the IPCC's goal of keeping global warming at no higher than 1.5°C, from then onwards 'there must be "no new oil and gas fields approved for development; no new coal mines or mine extensions."' Yet, although not one new fossil fuel project could happen if the world were to avert warming by more than 1.5°C, the following year saw a 'fossil fuel frenzy.' In 2022, 119 oil pipelines, 477 gas pipelines, over 300 liquified gas terminals, and 432 new mines were either planned or already under construction. Then, later that year – in part fuelled (pun intended!) by Russia's invasion of Ukraine – what Malm and Carton have called a fossil fuel 'bonanza of all times' took place. The massive record profits – hitting 'unprecedented stratospheric heights' – made that year by oil giants like ExxonMobil, Chevron, Shell, BP, Total and Saudi Aramco were in part dished out to shareholders. However, a huge proportion of those profits were used in a typical capitalist way: put into even more new fossil fuel projects in order to accumulate even more capital. That's why blockadia and its various methods remain important tools in the on-going struggle to build a better world.[8]

Keeping it in the ground

Another way of keeping fossil fuels 'in the ground' – where most of the remaining reserves need to be kept – is by pushing for measures which make fossil fuels a 'stranded asset', while at the same time improving the lives of the majority of the population. While it's true that significant reforms to, and limitations on, the 'normal' functioning of capitalism are often overturned as soon as the capitalist class and their defenders once again feel strong enough to do so, that is no argument against fighting for them and their implementation. Even reforms that in no way threaten the capitalist system can nonetheless improve the lives of ordinary people – and, by doing so, encourage more to join the struggle for even better reforms. Whilst fighting for transitional reforms, that truly weaken the capitalist monster, also helps to build and strengthen the forces needed to eventually bring about the creation of a better world – one that is socially-just and ecologically-sustainable.

When it comes to the fossil fuel corporations, there are two areas that, if tackled the right way, could seriously weaken them and slowdown their climate – and planet-wrecking activities: taxation and subsidies. As regards taxation, probably the most useful proposal was made in 2009 by James Hansen, one of the world's leading climate scientists. This was his idea of a fee-and dividend tax on the fossil fuel polluters, and was set out in his appropriately-titled book, *Storms of my Grandchildren*, in which he concluded that most of the fossil fuels still remaining 'must be left in the ground', and that a 'global phaseout of fossil fuel carbon dioxide emissions is a stringent requirement.' According to Alan Thornett, fee-and-dividend 'seeks to provide an effective framework for a very big reduction in fossil fuel emissions, here and now whilst capitalism still exists.' As Hansen explained, 'a rising carbon price' will make it 'economically senseless to go after every last drop of oil and gas.' Essentially, fee-and-dividend is a uniform tax that is, at least in part, based on calculations of the true cost of fossil fuels to society. This fee or tax is imposed on fossil fuels companies at the point of production, based on each ton of CO_2 contained in the fuel. The general public pays no extra tax – but obviously the price of goods they buy will increase, depending on how much fossil fuel is used in the production of those goods. To off-set that, '100 per cent of the money collected from the fossil fuel companies at the mine or well is distributed uniformly to the public.' As fossil fuels become more expensive, people would use their dividend to purchase cheaper renewable energy, and goods produced and transported by using such greener energy. Basing his calculations on the US in 2017, Hansen estimated that a fee of $115 per ton of CO_2 would raise $670 billion – with the resulting dividend being 'close to $3,000 per year' for each adult (children would get half-shares), with families with two or more children getting $8-9,000 per year. Given that the wealthiest people are disproportionately the greatest CO_2 emitters because of their lifestyles, such families would be paying out a lot more than the $9,000 dividend they'd get back. Such a progressive form of taxation means Hansen's proposal 'provides an exit strategy from fossil fuel based on the principle of making the polluters pay.'[9]

Such an approach was necessary, according to Hansen, because he believed the new so-called targets – based on improving the Kyoto Protocol – then being announced for reducing CO_2 emissions by 80 per cent by 2050 did not 'have a prayer of achieving that result' and that governments were deceiving the people when they said such goals were achievable. This is something most recently explained by Malm and Carton, who show that planned 'overshoot' – i.e., going beyond any set targets – was always part of capitalism's way to have its cake and eat it. For Hansen, simply slowing down the rate of CO_2 emissions was 'far from enough' if climate collapse was to be avoided. What was necessary was to *reduce* CO_2 concentrations back to 350ppm by 2100 – and that could only be done by leaving much of the remaining fossil fuels in the ground.[10]

Although Hansen isn't calling for the end of capitalism, his proposal does offer a fairly rapid way of coming off fossil fuels, and has 'the potential for mobilising the kind of popular support that would be necessary for such a rapid change.' Thus, his fee-and-dividend suggestion is arguably 'strongly in the transitional framework' – especially if it's combined with removing, or at least significantly reducing, the massive subsidies currently given to fossil fuel companies. That money could then be put into a rapid expansion of cheaper renewable energy schemes and a massive programme of home insulation. Such a combined approach would, quite literally, price fossil fuels out of the market. Such reforms – though obviously still allowing capitalism to continue existing – would be an excellent advance on the road towards building a 'carbon-free economy.' Winning such radical transitional reforms will not be easy – but is possible. If we really do want a better world, then those are precisely the kind of changes we need to be pushing – and pushing hard – for in the here and now.[11]

It's not surprising that Hansen's scheme is supported by some leading ecosocialists – what *is* surprising is that even more ecosocialists haven't been more supportive. However, in the US, Hansen's fee-and-dividend proposals quickly gained the support of John Bellamy Foster. In 2010, along with his co-authors of *The Ecological Rift*, he specifically made the point that Hansen's proposal was 'especially noteworthy.' In the appropriately-named section, 'Transition Strategies', after having pointed out that 'not all carbon taxes…are radical measures', the authors nonetheless conclude that Hansen's 'emergency strategy, with its monthly dividends, is designed to keep carbon in the ground and at the same time appeal to the general public.' What is particularly telling is that they emphasize how Hansen's scheme 'explicitly circumvents both the market and state power in order to block those who desire to subvert the process.'[12]

A clearer exposition of what transitional reforms look like, and are all about, could not be had. As they explain, this scheme offers the hope of building 'a mass popular constituency for combating climate change by promoting social redistribution of wealth toward those with smaller carbon footprints (the greater part of the population).' In other words, 'it would help to mobilize the population, particularly those at the bottom of society, in favor of a climate revolution.' Such a scheme not only would help us build a better world in the immediate future, but would also weaken the enemies of humanity whilst emboldening the popular movements needed to bring about a true ecological and social revolution.[13]

Foster returned to this issue in a 2013 paper, commending Hansen's fee and dividend scheme because 'it is directly aimed at making the fossil fuel companies…pay, while increasing the price of carbon to decrease consumption in every nook and cranny of the economy.' By doing so, it would help 'prevent the world from crossing a catastrophic climate tipping point.' Whilst it doesn't tackle capitalism's drive to constantly augment its accumulation of capital – and thus doesn't present a solution to the long-term aim of re-placing capitalism to re-establish 'climate or ecological stabilisation' – it is nonetheless 'objectively revolutionary' in that it 'would make clear…the class nature of carbon footprints and the increasing destruction of the planet as a place of human habitation.' By bringing Hansen's scheme to the attention of ecosocialists and the wider climate movement, ecosocialists like Thornett see Foster's work as 'an important contribution to the struggle against climate change.'[14]

More recently, in 2018, Anders Fremstad and Mark Paul also gave strong support to a version of a fee and dividend scheme proposed by the People's Policy Project, arguing that such a scheme 'could dramatically cut emissions while redistributing income' and thus help fight climate change and reduce inequality. Their conclusion was that 'in a time of growing inequality, persistent poverty, and woefully insufficient social insurance programs, a carbon dividend provides a climate solution that is both environmentally and socially sustainable.' Clearly, such radical proposals, 'though ostensibly transitional strategies', also 'present the issue of revolutionary change' as movements strong enough to win such reforms could also become strong enough 'to implement a full-scale social ecological revolution.'[15]

Green New Deals

Another potential way of weakening the capitalist monster is by pushing for the implementation of Green New Deals (GNDs). The idea for GNDs has been around in several countries for over a decade but, according to Jonathan Neale, 'the decisive moment came in 2017' when, in the US, two leading Democrat politicians and campaigners – Alexandria Ocasio-Cortez and Bernie Sanders – 'decided to back a Green New Deal.' By 2019, GNDs were being increasingly supported by various climate and labour movements around the world; during and after the pandemic, the call for GNDs spread even further. Although the demand for climate jobs – to 'solve' the climate crisis – was a central part of the GNDs being called for, they were also focussed on solving other crises inflicted by capitalism, such as job insecurity, and social and economic inequalities. As Naomi Klein explained, the idea of a GND is a simple one: 'in the process of transforming the infrastructure of our societies at the speed and scale that [climate] scientists have called for, humanity has a once-in-a-century chance to fix an economic model that is failing the majority of people on multiple fronts.' That 'economic model', of course, is capitalism – a monster that is both increasingly trashing planet Earth, and inflicting ever-more obscene inequalities on people living in the global North and, especially, those living in the global South. Most supporters of GNDs envisage schemes on a par with the New Deal which helped get the US out of the worst impacts of the Great Depression of the 1930s, and the massive efforts so many states put into the war against Nazi Germany after 1939. The hope is that such massive state expenditure would – as well as creating millions of climate jobs around the world – also transform economies in ways 'both to protect and to regenerate the planet's life support systems and to respect and sustain the people who depend on them.'[16]

In the US, in February 2019, Alexandria Ocasio-Cortez and Democrat Senator Ed Markey presented a Resolution to Congress, calling for the Federal Government to implement a Green New Deal with the aim of making the USA 'net carbon-neutral in ten years.' This GND proposal called for 'huge strides in reducing the USA's reliance on oil, gas and coal and its replacement by clean, renewable and zero-emission energy sources.' For achieving those aims, the Resolution called for 100 per cent of US power to be provided by renewable and zero-emission energy sources; the insulation of all existing buildings, with new buildings to be energy-efficient; and removing greenhouse gases from the atmosphere by restoring natural ecosystems. Importantly, as regards building popular support for such a programme, it also had plans for a just transition for all communities and workers, and for eliminating poverty. Unsurprisingly, Trump dismissed the GND Resolution 'as "socialist and therefore un-American."' Similar attempts to get GNDs accepted were also made in Canada, Europe, the UK, and Australia.[17]

On one level, all the main GND proposals have one important feature in common: they all attempt to make the main polluters pay the costs of the damages they've knowingly caused, and to cough up the bulk of the money needed for the task of keeping global boiling as close as possible to 1.5°C. Many climate and environmental activists – and ecosocialists – have thus correctly argued that, in view of the mounting evidence of climate crisis, all GND plans 'must push the boundaries of the situation' by becoming even more radical in order to build the broad alliance needed to defend people and planet. Significantly, achieving a rapid move to a new energy system – based in particular on solar and wind – and a just transition from carbon-based jobs to jobs based on renewable energy, is too important to be left to parties and governments. This was dramatically underlined by what happened in Britain, following Keir Starmer's election as Leader of the Labour Party in 2020. Under Jeremy Corbyn, the Labour Party had adopted a radical GND agenda – seen by some as 'greener' than that put forward by the Green Party in the 2019 general election. Although in 2021 Starmer had announced his own 'Green Prosperity Plan', in February 2024, he publicly reneged on his pledge to invest £28bn every year, over ten years, to enable the UK to transition to clean energy. Instead, he revealed it would be reduced to £4.7bn – and he made that announcement on the very same day that the EU's climate agency revealed that the average global temperature was 'continuing to accelerate at a highly dangerous rate', and that, *for the very first time,* it had exceeded the 'agreed' maximum of

1.5°C above preindustrial levels 'every day for a whole year.' The previous month – January 2024 – had just been confirmed as the hottest January ever recorded; and, as has been seen, the year continued with temperature records being breached almost every month. Starmer's decision was described by veteran ecosocialist Alan Thornett as a 'miserable betrayal' of the planet; while Barry Gardiner (who'd been Corbyn's Shadow Minister for the Environment) was even more outspoken, publicly condemning it as 'economically illiterate, ecologically irresponsible, and politically inept.' As many environmentalists argued after the decision was announced, it is: 'only governmental action, based on strong public investment and with political will behind it, that can generate the momentum needed to carry through the green transition and tackle the energy crisis by providing cheap and sustainable renewable energy.[18]

Nonetheless, as the climate and ecological crises continue to worsen, the struggle for GNDs needs to be made even more determined – especially within the labour and climate movements. As regards the former group, GNDs will have to map out, in clear ways, how stopping further global boiling can be achieved via a 'socially just transition to renewables.' The climate and environmental crises are crucial issues confronting everyone, including those in the labour movement, and so cannot be treated as 'bolton' issues to the more traditional labour movement concerns. The more we allow capitalist corporations to degrade the planet, the more difficult it will be to achieve social and economic justice for the majority of people – in the global North as well as in the global South. The increasing urgency for putting GNDs into operation as quickly as possible should be abundantly clear by now: 'the temperature is still rising by record margins, the deserts are still expanding, the forests are still burning, the ice is still disappearing, the sea is still rising and filling up with plastic, and weather events are getting more severe, and the other species with which we share the planet are being driven extinct at an ever-increasing rate.' Thus, the GNDs we need are those which do not copy the capitalist drive for growth – instead, such GNDs should be based on expanding public services and welfare by such transitional reforms as Universal Basic Services; a just transition for people during the move to renewables and more useful forms of production; and a big reduction in working hours.[19]

Living more lightly

Elinor Ostrom was just one environmentalist who advocated 'an approach of living lightly on the Earth, using fewer resources and using them with care' but who was also 'well aware that for policies to gain support they had to improve people's lives.' Both the environmental and labour movements are generally very good at pointing out what's wrong with the world today – but not so good at creating programmes and narratives about the future that are appealing enough to swing over large numbers of people to supporting significant transformations in our relationships with each other and planet Earth. Three areas that can help map out a better and more sustainable world are those connected – for want of a better term – with 'degrowth'; the question of where the bulk of our food comes from; and, most controversially of all, consideration of the issue of population. In some of those issues, the question of individual life-style choices arises – and is not something which should be down-played. However, in the three areas discussed below – and in all other issues which will arise in the struggle against capitalism and the construction of post-capitalist societies – one of the most vital things will be democracy. Not the old-style top-down politics based on occasional elections, but a grass-roots, inclusive and bottom-up democracy. In the UK, Extinction Rebellion is just one group pushing for this via their calls for upgrading democracy and Citizens' Assemblies.[20]

Post-capitalist thriving

In 1992, Ernest Mandel pointed out that 'Today we have become aware, with much delay, that dangers to the earth's non-renewable resources, and to the natural environment of human civilization and human life, also entail that the consumption of material goods and services cannot grow in an unlimited way.' Mandel was one of an increasing number of people who recognized that a society and economy that meets the true needs of both humans and nature will value different 'commodities' from those 'on offer' from capitalism: such as better public services and greater leisure time. Since then, others have increasingly drawn attention to the dangers associated with unlimited growth. At the start of the twenty-first century, a radical 'degrowth' movement began to emerge – initially in France, but soon spreading across the globe. In 2002, Joel Kovel published *The Enemy of Nature*, and then, with Michael Löwy, a shorter *Ecosocialist*

Manifesto. Sadly, much of the left dismissed such growing concerns as 'middle-class environmentalism' and even as a distraction from the 'more important' struggle for socialism. In 2018, the economist Kate Raworth suggested putting 'GDP growth' aside and, instead, posed this fundamental question: 'what enables human beings to thrive?' As she pointed out, since 1950 – roughly when the economic era known as the 'Great Acceleration' began to take off – 'real World GDP increased sevenfold.' One mounting result of this growth has been 'an accompanying surge in ecological impacts, from the build-up of greenhouse gases in the atmosphere to ocean acidification and biodiversity loss.' Clearly, such exponential growth has put Earth's life-support system under increasing pressure – but, as Raworth suggests, the crucial question is: 'just how much pressure can it take before the very life-giving systems that sustain us start to break down?' In place of ever-more GDP 'growth' – vital to the functioning of capitalism, but increasingly deadly for people and planet – Raworth suggests measures by which people can thrive in balance with the natural world.[21]

More recently, writers such as Giorgos Kallis, Jason Hickel and Kate Soper have also argued for the need to get off the 'more-growth' treadmill. The alternative approach advocated – sometimes called 'degrowth', or a 'post-growth living' that offers people an 'alternative hedonism' – is 'to purposely slow things down in order to minimize harm to humans and earth systems.' For Kallis and his co-authors, 'caring and community solidarity are vital principles of degrowth societies, and engines for moving in more equitable and sustainable directions.' Fundamentally, 'degrowth' is about moving away from the 'resource-intensive and ecologically damaging aspects of current economies.' While Jason Hickel has argued for breaking with the assumption that even high-income countries need constant growth and, instead, to realize that 'we can improve human well-being without having to expand the economy.' This won't be easy, as even proponents of what's called 'green growth' are willing to back yet more growth, on the assumption that some future technology will allow us to put right the harms that are currently being done to the planet. 'Degrowth' is a concept that sounds negative – and, of course, defenders of capitalism and the importance of constantly increasing GDP argue that it is 'growth' which has been 'responsible for the extraordinary improvements in welfare and life expectancy that we've witnessed over the past few centuries.' However, as Hickel points out, historians and scientists are now able to show that 'it's not growth itself that matters – what matters is what we are producing, whether people have access to essential goods and services, and how income is distributed.' The crucial point – from an ecological point of view – is that once societies have reached a certain level of economic development, 'more GDP isn't necessary for improving human welfare at all.'[22]

Kate Soper, too, is concerned about where constantly-increasing GDP and consumption is taking us. She calls for fundamental changes in the 'pattern of consumption in affluent societies', so that we can build 'a more egalitarian and sustainable global order.' To do that, she argues that what's needed – mainly in the global North – to achieve that is 'a cultural revolution in thinking about prosperity, and the abandonment of growth-driven consumerism.' Soper thus advocates an 'alternative hedonism' that, instead of seeing 'the needed changes in consumption as a form of sacrifice and loss of pleasure', we should take enjoyment and pleasure from 'having more time, doing more things for oneself, travelling more slowly and consuming less stuff.' As she points out, constant increases in production and consumption – i.e., the capitalist way of life – is not 'just environmentally disastrous but also in many respects unpleasurable, self-denying and too puritanically fixated on work and money-making.' As is increasingly apparent, many are condemned under consumer capitalism to work long hours, undertaking often very boring tasks, to produce largely-useless stuff that does little to create real human happiness. Or, as Foster and others have put it: 'Unnecessary, wasteful goods are produced by useless toil to enhance purely economic values at the expense of the environment.'[23]

Though 'degrowth' can be a difficult concept to argue – especially to those who already have difficulty in accessing goods – it is nonetheless too-important a debate to sidestep. As Alan Thornett argued in 2023, the climate and ecological crises mean that we have to face up to the 'harsh reality' that constant growth is driving ecological breakdown and 'is unsustainable on a finite planet'; thus, the left 'must embrace degrowth as an alternative to capitalism's endless pursuit of growth.' Quite simply, the evidence is absolutely clear: to argue that constantly 'growing' economies is the *only* option for humans is 'a ludicrous proposition' on a finite planet, and needs to be challenged. However, constant economic growth is the only option for capitalism – and, for many decades, most of the left have also embraced the idea of economic growth and productivism as being the only way to improve conditions of life for the majority of the population. The 'harsh reality' is that 'exponential economic growth' is the biggest cause of global boiling and ecological destruction, because 'growing economies require constant inputs of resources, and constantly transform land, water, and air shared with other species… Growing economies also mean growing waste and pollution.' This is why current growth patterns are 'a direct threat to the future of life on the planet' – and why ecosocialists propose 'degrowth' as a strategic alternative to capitalism's destructive addiction to constant growth: an alternative that is based on a planned and socially-just degrowth of the size of the economies of the richest countries. A 'degrowth' society would reduce or even abolish certain products, whilst subsidising and expanding those that could be produced in harmony with ecosystems and the non-human species living on this planet. However, Jonathan Neale – who previously supported the idea of 'degrowth' – wrote an article in 2023 ('The "eco" in ecosocialism must mean climate, or we are lost.'), in which he called for dropping the call for 'degrowth', arguing that 'degrowth will not halt climate change', and pointing out that even reducing global GDP by 50 per cent 'in the next twenty years' would not be enough to stop global boiling if the world continued to burn fossil fuels. He then went on to argue, conversely, that a 50 per cent *increase* in global GDP would be ok, as long as 'we stop all burning of fossil fuels.' Not surprisingly, this shift was criticized by several ecosocialists who, for instance, pointed out that the strong link between GDP and GHG emissions was shown during the COVID-19 pandemic: 'According to Statista GDP fell by 3.4 during the lockdowns and GHG emission by even more, by a whopping 4.6 per cent.' Of course, such unplanned and socially-unjust 'degrowth' is not what those who support a progressive 'degrowth' are arguing for: on the contrary, what is needed is a 'degrowth' transition that ensures better lives for the majority. Furthermore, tackling fossil fuel use and environmentally-destructive production are not mutually exclusive campaigns – *both* are essential in achieving the 'System Change' we need to see.[24]

Ultimately, if we fail to grasp the nettle of 'degrowth' and, instead, carry on with business-as-usual, we will be 'sleepwalking towards ecological and societal breakdown.' Essentially, this is because it is now obvious to many that – as noted by André Gorz –

maintaining the delicate balance of Earth's life-support systems is incompatible 'with the survival of the capitalist system.' The bottom line with 'degrowth' is that it's not about making people 'poorer' – but about having an ecologically-sustainable shift from constant growth to building societies based on 'useful production, democracy, economic redistribution, and well-being.' The call for 'degrowth' is thus an essential part of the ecosocialist project – both today in the struggle against capitalism, and in the post-capitalist future – to construct ecologically-resilient and socially-just societies that produce and consume differently, in order to reduce our impacts on the planet; and that allow people to live socially-rich and meaningful lives. One of those who has argued that 'degrowth' would actually make living better and happier for people is Ernest Mandel, who was for many years the Fourth International's principal theoretician: 'The continual accumulation of more and more goods (with declining 'marginal utility') is by no means a universal or even predominant feature of human behaviour. The development of talents and inclinations for their own sake; the protection of health and life' care for children; the development of rich social relations as a prerequisite of mental stability and happiness – all these become major motivations once basic material needs have been satisfied.' While capitalist globalization has undoubtedly created a whole swathe of 'personal needs, beyond solely consumerist aspirations', the important point is that capitalism 'cannot satisfy these aspirations for more than a small minority', but its constantly-increasing production of 'stuff', in order to increase capital accumulation, is now destroying the planet's life-support systems to such an extent that the livelihoods and even lives of millions of individuals are already at serious risk. As the nineteenth-century revolutionary Auguste Blanqui noted so long ago, capitalism effectively 'exterminates' the individual because it 'cares for individuals as much as it does for earthworms.'[25]

Palate or planet?

There is now overwhelming evidence that one of the biggest drivers of both the climate and the ecological crises is capitalist agriculture, which treats the natural world as a relatively insignificant part of its productive process. Whilst no agricultural production can fail to have some impacts on nature, those of global capitalism's highly-industrialized agriculture are so negative because – like all other aspects of capitalism – it is geared to producing food for profit rather than for use. Essentially, capitalist agriculture is environmentally irrational and unsustainable. This is the point made by documentaries such as *Eating Our Way to Extinction* and *Eating for Tomorrow*.[26]

In 2015, Fred Magdoff examined the negative impacts on ecosystems of intensive capitalist agriculture in the US. Many of his findings are applicable globally: these include biodiversity loss, as native plant species are eradicated in order to grow crops for profit; while the loss of habitat for a wide range of species means that natural control mechanisms are also lost. As he concluded: 'All of the common decisions and practices in the agricultural system... [are rational] only from the very narrow perspective of trying to make profits within a capitalist system.'[27]

Nowhere is this more obvious than in relation to animal agriculture. Apart from its high emissions of greenhouse gases – which total more than all forms of transport, including flying, combined – one of the most dramatic is related to land use, deforestation, and biodiversity/species loss. In fact, animal agriculture is the world's largest single driver of deforestation. This is particularly marked in the Amazonian rainforest – often described as the 'lungs' of the planet – which is an essential part of Earth's ecological equilibrium. Almost 70 per cent of the cleared Amazon rainforest is used for grazing cattle for the international beef and leather industries. As Ian Angus pointed out in 2016: 'Most of the land now being converted to agriculture was formerly tropical forest.' Essentially, Brazil's tropical rain forests are disappearing at an alarming rate, to provide short-term grazing for cattle to produce quick profits for big landowners. This destruction has continued apace, with the result that, globally, it's estimated that in the past 50 years or so, at least one third of the world's woodland has been destroyed – much, if not most, of it for intensive meat and dairy agriculture. Study after study has shown that both those industries need, at the very least, to be drastically reduced, if we are to create sustainable agro-systems that work for people and planet instead of for corporate profit.[28]

However, as well as being so destructive of natural habitats and ecosystems – in part, because of its intensive use of pesticides, herbicides and artificial fertilizers, and its depletion of freshwater sources – capitalist meat and dairy agriculture is also highly irrational and inefficient when it comes to feeding people. Again, various studies have shown that by shifting massively away from meat and dairy production, the world could adequately feed a population much larger than today's 8+ billion. When it comes to producing proteins for human consumption, 100kgs of plant protein is needed to produce 9kgs of beef protein or 31kgs of milk protein. Another way of looking at this question is that 10 hectares of land yields enough soya to feed 61 people, or enough grains to feed 24 people. But the same land would only produce enough maize to feed ten people – and only enough meat to feed two people! As Thornett has commented, meat and dairy animals 'consume vast quantities of corn, maize and soy that could otherwise be eaten, far more effectively, by the human population including the planet's billions of hungry people.'[29]

Currently, over 50 per cent of all crops grown is fed to farmed animals – around 80 per cent of which are now 'intensively' farmed. The big capitalist agri-businesses now require roughly 70 per cent of the world's land – either as grazing for animals (increasingly becoming a rare thing), or for growing crops for animal feed. Just one hamburger made using Costa Rican beef results in the destruction of one large tree; 50 saplings; almost 30 different species of seedlings; and hundreds of species of insects, mosses, fungi and micro-organisms. As the saying goes: 'You are what you eat'! In all, it's estimated that around 60 per cent of global biodiversity loss is directly attributed to capitalist agriculture.

Yet recent studies have shown that a massive shift away from meat and dairy, and towards a mainly plant-based diet, would allow much agricultural land to be restored to nature. In 2018, a study by the LEAP (Livestock, Environment and People) project at Oxford University showed that animal farming took up 83 per cent of farmland and was responsible for 60 per cent of all agricultural GHG emissions, but only provided 18 per cent of the calories needed by humans. It also found that 'even the very lowest impact meat and dairy products still cause much more environmental harm than the least sustainable vegetable and cereal growing.' It concluded that: 'without meat and dairy consumption, global farmland use could be reduced by more than 75%... – and still feed the world.' As George Monbiot has pointed out, all the evidence now clearly shows that the crucial shift needed is from an animal – to a plant-based diet: 'We can cut our consumption of everything else almost to zero and still we will drive living systems to collapse, unless we change our diets.' Those 2018 findings were confirmed by LEAP researchers in 2023, showing that vegan diets 'could significantly reduce emissions, water pollution, and land use.' While two other studies, published in 2021 and 2023, concluded that: 'as of 2015, food-system emissions accounted for roughly 34% of global emissions', and that 'over 70% of freshwater use is attributed to agriculture.' As regards the UK, moving to a plant-based diet would release 75 per cent of UK farmland, which would allow 50 per cent of the UK's land-surface to be restored to nature.[30]

There are, of course, other reasons for supporting a massive shift away from meat and dairy, apart from concerns about the natural world. Those relate to ethical questions about animal welfare and slaughter, and about the implications of meat and dairy consumption for human health. Over 70 billion farmed animals are slaughtered every year – and, as a recent survey by the vegan charity Viva! showed, 'over 80 per cent of all farmed animals are condemned to factory farms – filth, stench, overcrowding, cannibalism, mutilations and barred cages.' Undercover campaigners have found 'barbaric conditions' and 'major breaches of UK animal welfare standards' even on farms which had been approved by the Red Tractor scheme – which is owned and funded by the British farming and food industry. One such example, of a pig farm which supplied Tesco, became the subject of a documentary, *Hogwood: A Modern Horror Story*, which can be streamed. Such animal cruelty was highlighted earlier by Fred Magdoff and Chris Williams in 2017: 'Raising farm animals using industrial farming practices is cruel to the animals... The capitalist imperative to raise animals as quickly as possible in order to reap the highest profits results in inhumane crowding conditions for the animals.' Such conditions are particularly associated with what are 'mega-farms' – called 'Concentrated Animal Feeding Operations (CAFOs) in the US. There are now over 800 such 'mega-farms' in the UK – up by over 25 per cent since 2011. Many of these use 'no-graze' systems, with firms such as Müller and Arla increasingly sourcing their dairy products from such mega-farms. According to a BBC report in August 2024, there are 802 'larger-scale beef and dairy farms' in the UK. However, one farmer who 'continuously houses up to 1000 dairy cows on one site', believed there were more 'no-graze' farms than official statistics showed, because 'A lot of people are doing it but they don't always say it because they know it causes a backlash.'[31]

As well as being damaging to the environment and unkind to farmed animals, intensive animal agriculture is also damaging to human health. In particular, the routine use of antibiotics and growth hormones on farmed animals, in order to maintain and increase 'yields' is almost certainly the main cause of increasing antibiotic resistance. According to a report by the World Health Organization in 2017, the routine use of antibiotics in animal agriculture, 'to promote growth and prevent disease in healthy animals' was resulting in some types of bacteria that cause serious infections in humans developing resistance to antibiotics. As the report warned: 'A lack of effective antibiotics is as serious a security threat as a sudden and deadly disease outbreak.' And, as has already been seen, the destruction of huge swathes of the natural world by capitalist agri-business corporations is one of the main reasons for the increasing frequency of zoonotic viruses such as COVID-19. Finally, there is increasing evidence that a (mainly) plant-based diet is significantly healthier for humans than one which involves the consumption of meat and dairy – this applies to a number of human diseases, including heart disease and cancer. But if you're being to consider a switch to a more plant-based diet – whether for the planet, the animals, or your own health – be prepared for questions from family or friends along the lines of: 'But where will you get your protein from?' Apart from the fact that all protein comes, initially, from plants, a good response is to ask them whether they've looked closely at a gorilla and been worried that they weren't getting enough protein for their muscles! Gorillas are vegan – and we share 98 per cent of our DNA with them. For those who think they could never make such as transition, try watching the brilliant documentary film *I Could Never Go Vegan* for a range of reasons why maybe you should at least *begin* to travel that way – if you want a world that is kinder to the planet, all Earthlings and ourselves.[32]

The 'p' word

This next topic is a political 'hot potato' – however, the 'p' doesn't stand for 'potato', but for 'population.' This is a very contentious issue, even – or maybe especially – amongst ecosocialists. First of all, it's important to establish some crucial points: human population levels are *not* the main drivers of the climate and ecological/biodiversity problems; nor are they the cause of hunger, inequality and poverty. By far the main cause of all those problems is capitalism – thus 'any attempt to link the rising global population to anti-foreigner or anti-immigrant racism is reactionary and should be condemned.' As regards the climate crisis alone, while the highest birthrates are in the global South, the total carbon footprint of those countries is far exceeded by the wealthier countries in the global North. Secondly, also totally unacceptable – including for the reasons outlined above – are any reactionary eugenics-based neo-Malthusian arguments, as are coercive state-sponsored 'population reduction' programmes such as forced sterilization.[33]

However, even after capitalism has been replaced by societies which are kinder to both people and planet, it's likely that long-term trends in total global population might still be matters for public debate. To carry the point *ad absurdum*, Earth has a finite amount of land – which means there will, logically speaking, eventually come a point where continuous population growth could lead to a situation in which, literally, there won't be space for everyone to find standing room, never mind space for growing food and building shelters for everyone. Getting back to reality, in 1950, the world population was around 2.5 billion: today, it stands at around 8 billion. In other words, it has more than tripled in 75 years, with a fairly stable increase of over 70 million a year. While it is true that the global birthrate is now falling, and that the UN believes that total population will stabilize and even start to decline from about 2050 onwards, it nonetheless predicts that world population by then will have reached just over 11 billion. However, almost 'half of the current global population is under twenty-five' (with an obvious potential for significant further growth), while 'the per capita consumption of food, water, and manufactured goods is increasing even faster than the population itself.' Logic would suggest that, because the Earth is finite, there is an eventual limit to the ecological impacts the planet can sustain – and that the number of people, their local and global environmental impacts, and their levels and types of consumption are likely to have significant effects on the Earth's ecosystems and their ability to continue to support life. Even living at a subsistence level, the environmental footprint – as opposed to carbon footprint – of each person necessarily includes the food, water and land needed to survive: which often necessitates the destruction of further parts of the natural world. This is without factoring in that, quite rightly, many in the global South – as well as the poor in the global North – want to enjoy a higher standard of living than they currently have.[34]

Ecosocialists such as Ian Angus and Derek Wall, however, have argued against seeing world population totals as any sort of significant problem – sometimes out of understandable fear of giving some credence to reactionary rightwing neo-Malthusian 'arguments.' As Derek Wall has pointed out, some 'population lobbyists…increasingly target not just human numbers but human movement.' Wall tends to see those who advocate empowerment of women and education as ways to 'reduce poverty and environmental destruction' as offering 'flawed' and even 'very problematic' arguments. In contrast, he has made the point that: 'We consume too many resources not because population is rising but because the capitalist economy depends on us consuming more.' While, in a debate with Alan Thornett, Ian Angus and Simon Butler rightly argue that the whole question of population growth and 'humanity's complex relationship with nature' cannot be reduced 'to simple numbers.' In support of their view, they reference the Mexican feminist and human rights activist Lourdes Arizpe who, in 2014, made the point that the whole issue of population – seen as numbers of human bodies – is a much wider one: 'It is what these bodies do, what they extract and give back to the environment, what use they make of land, trees, and water, and what impact their commerce and industry have on their social and ecological systems, that are crucial.'[35]

Nonetheless, according to Thornett, one way of approaching the question of population is by focussing on empowering all women, both in the global South and in the global North, 'to control their own fertility providing it is based entirely on a woman's right to choose, and full social justice, whilst rejecting any and all forms of compulsion or coercion.' Such an approach does not, as suggested by some, target the women of the global South and 'blame' them for the climate and ecological crises: what it does target are 'the appalling conditions women in the Global South are forced to face, and the unmet need for reproductive services that they are forced to suffer.' In addition, part of such an empowerment programme – contrary to 'turbo-charging' reactionary views – would also help more women gain access to education and jobs, lifting them out of poverty, and giving them independence from patriarchal and conservative pressures. As regards the global South as a whole, this is why ecosocialists argue strongly for climate justice, 'to ensure that such countries get a massive transfer of wealth from the rich countries of the global North to help them both get out of poverty and transition to renewable energy.' All these ongoing discussions suggest that population is likely to remain a debatable issue even in post-capitalist societies.[36]

Moving forwards to a better world

Having had a comprehensive 'look at our history' – to reference Ruby Turner once again! – it should be abundantly clear that, if the Earth is to live and *all* its inhabitants thrive, 'there has got to be a better way' of doing things. The good news is that there is. As with any problem, the first crucial step towards solving today's manifold problems and finding a better way, is to identify the cause of most of those problems we – and the planet – are facing. By far the main cause is not humans – or even industrialism – but capitalism itself.[37]

Obviously, winning transitional reforms like those suggested above – both under capitalism, and in the immediate post-capitalist future – in order to create a world in which humans live in lighter and kinder ways, is going to be a difficult task. However, as the long-time environmental campaigner Wendell Berry said in 2013, in an interview titled 'Confronting the Consequences of Runaway Capitalism': 'We don't have a right to ask whether we're going to succeed or not. The only question we have a right to ask is what's the right thing to do? What does this earth require of us if we want to continue to live on it?' So, what do we need to do right now, in order to start building that better world? On one level, doing what we can as individuals concerning the impacts of our lifestyle choices – on the planet and *all* its inhabitants – is by no means to be dismissed as merely 'moralistic' or 'well-meaning.' After all, if nothing else, would it really be *better* 'if people had no regard for their personal impact and just carried on eating more hamburgers, driving more miles, and taking more flights?' In fact, the chances are that someone who is aware of the impact of their individual actions on nature are much more likely to see the need also to take *collective* action to bring about wider changes than someone who doesn't care. Furthermore, for an individual's lifestyle to be significantly at odds with their publicly-professed philosophical and political beliefs would be hypocritical. Though perhaps it's not necessary for us to be exactly like 'the anarcho-syndicalist, who ruthlessly subjects [their] private life to the norms upon which [they want] to base the laws of a future social state.'[38]

However, winning transitional reforms will require the building of the broadest possible mass movement, consisting of all those prepared to struggle to save the planet: 'the environmental movements, the indigenous movements, peasant movements, and farmers movement, as well as trade unions and progressive political parties.' It is such a mass movement that will be able to make *all* the big capitalist polluters – not just the fossil fuel corporations – pay for a socially-just transition to renewable energy and an economic system that puts the goal of human thriving above the greed of a small minority. Part of that will necessarily involve winning wide popular support for a major global redistribution of wealth from the rich to the poor. Crucially, we need to achieve that *before* the multiple crises of capitalism cause disastrous climate, ecological and societal breakdown – because in such conditions, which would see an even greater growth of far-right populism and creeping fascism, it will be hugely more difficult to achieve the transitional reforms needed 'to save the planet from ecological destruction and create a post-capitalist, ecologically sustainable, society for the future.' How to build such a mass – *and* victorious – movement will be the focus of the next chapter.[39]

(*WE SAY WE WANT A*) REVOLUTION

*Sweating blood and filth with every pore from head to toe"
characterises not only the birth of capital but also its progress
in the world at every step, and thus capitalism prepares its own
downfall under ever more violent contortions and convulsions.*[1]
Rosa Luxemburg

I don't think John Lennon would have minded too much about my using parts of The Beatles' song 'Revolution' as the title for this chapter. As he'd pointed out in a 1971 interview with *Red Mole* – then the paper of the UK's revolutionary International Marxist Group – after the line 'But when you talk about destruction, don't you know that you can count me out', he'd then sung 'count me *in*.' While the prospects for bringing about a revolution may seem very slim at the moment, the multiple and inter-linked ecological, social, and political crises that we are facing, make it increasingly obvious that huge changes are now needed. As Walter Benjamin once defined them, revolutions are not so much the 'locomotives of history', but rather 'humanity reaching for the train's handbrake before it falls into the abyss.' Quite simply, 'we will not survive if we continue to allow [capitalist] corporations to occupy a central role in the economy.' Furthermore, 'the only thing worse than an unjust economic system is an unjust economic system when it implodes.' In April 2024, a report published in the journal *Nature*, showed that just how nasty imploding capitalism is – and will increasingly become. According to one study, by the Potsdam Institute for Climate Impact Research, the current level of global boiling will result in average world incomes falling 'by almost a fifth' by 2050 – 'independent of future emission choices.' Overall, the total costs of climate damage 'will be six times higher' than 'the mitigation costs required to limit global warming to 2°C.' According to that report, climate damages are 'already locked in at about \$38 trillion a year by 2049'; by the end of this century, 'the financial cost could hit twice what previous studies estimate.' Unsurprisingly, hardest hit will be 'the poorest areas and those least responsible for heating the atmosphere.' All of this is of particular concern to younger generations who, according to what Hansen et al. said in November 2023, have been 'handed a planet in decline.'[2]

It is such information – which now seems to come every week and, sometimes, every day – which explains why, as well as struggling for much-needed immediate reforms, an increasing number of people are coming to see that we also need to be working for a revolution. Because, ultimately, 'the whole system has to go [if we are] to save ourselves and our planet.' According to some, bringing about a revolution has arguably become more difficult since the fall of the 'communist' regimes of the Soviet Union and eastern Europe. Enzo Traverso, for instance, believes that 'the defeats of the revolutions of the twentieth century have had a long-term, cumulative effect' with the result that 'for the first time in two centuries, revolutionaries [have] no paradigms to draw on.' However, ecosocialists see the demise of those undemocratic – and often environmentally-destructive – regimes as having *liberated* the revolutionary left from something which, as well as acting as an inspiration, was also more than a bit of a millstone. Left revolutionaries, Marxists as well as anarchists, have learnt important lessons about the need for real social(ist) democracy – something Trotsky had belatedly urged from the mid-1920s, and which was later taken up by the Trotskyist Fourth International from the late 1930s onwards.[3]

Why 'half' is not enough

The previous chapter focused on some of the reforms needed, right now, to slow down and mitigate some of the worst of the increasingly-destructive impacts of capitalism on people and planet. Yet it is hugely important to go *beyond* reforms – including transitional reforms – as they are only half-way emergency measures that can always be rolled back, if the existing power structures and relationships are left in place. As the French revolutionary Saint-Just pointed out in 1793: 'Those who [only] make half a revolution dig their own graves', because the ruling elites can always return, with 'the specter of counterrevolution, [which is] the eternal danger that haunts all revolutions as soon as they emerge' always threatening to disfigure attempts to construct a better world. All revolutions always start with significant reforms – a sort of half-revolution – in which the 'old' that needs replacing coexists alongside the new world that is 'struggling to be born', as Gramsci put it. Nonetheless, there are those who – despite recognizing, and opposing, some of the worst excesses of capitalism – still cling to the hope that it will be possible to 'green' capitalism: if not now, then at some point in what's increasingly becoming a very 'interesting' future. The sad reality is that less ambitious reforms, designed to allow the capitalist system to carry on operating, while they interfere (at least for a time) 'with the operations of individual capitalists, never go so far as to threaten the system as a whole.' Consequently, as Paul Sweezy has argued, long before the system itself is seriously threatened, such constraints are thrown off, as 'the capitalist class, including the state which it controls, mobilizes its defenses to repulse environmental-protection measures perceived as dangerously extreme.'[4]

One to explain, convincingly, why 'green capitalism' can't work is Daniel Tanuro who concludes that: 'Capital and nature do not speak the same language, and in the words of political economist Joel Kovel, one is necessarily the "enemy" of the other.' To ensure that we can create a real alternative to capitalism, 'one must go to the root of things.' In the end, merely restricting some aspects of their operation – though certainly important in the short term – will not be sufficient. Ultimately, the answer lies in carrying out a thorough-going revolution, or 'great transformation', in both economic and social terms. Or, as Extinction Rebellion has called for since its first appearance: 'System Change'. Nonetheless, the struggle for transitional reforms like those mentioned earlier – while only 'half a revolution' – can help us get to 'the root of things' by building the strength we will need to bring about a full revolution and so create a better world. This is because such reforms help us weaken the capitalist system – and make sure we struggle under more, rather than less, favourable circumstances. That's why we need, in the here and now, to 'slow capitalism's ecocidal drive as much as possible and to reverse it where we can.' However, fighting for reforms – even transitional reforms – and fighting to build an ecosocialist future, are not separate activities: 'they are aspects of one integrated process.' Defending human survival has to be a – though not the only – central aim; and, to put it bluntly, if we can't prevent a new oil or gas project, or win at least one significant reform, 'how can we imagine that we're actually going to over to overthrow capitalism?' As Edward Thompson said: 'The end of politics is to act, and to act *with effect*.'[5]

The whole idea of such transitional demands is very much associated with Leon Trotsky. In September 1938, after the rise of fascism, and with Europe on the cusp of the Second World War, he completed the final version of *The Transitional Program for Socialist Revolution* – this became the programmatic document for the new Fourth International, which was founded later that month. Trotsky saw such transitional demands as helping prepare the way to revolution – half-way bridges between 'minimum' demands which could be granted without significantly weakening the capitalist system, and 'maximum' demands which were, in fact, openly revolutionary. However, in times of crisis, 'the distinction between minimum and maximum demand would then collapse: the former would become a *transitional* demand.' In a phrase that's hugely applicable to the twenty-first century, he noted that the various crises of the capitalist system in the 1930s were inflicting 'ever heavier deprivations and sufferings upon the masses.'[6]

Trotsky was at pains to explain the significance of transitional demands during capitalist crises: 'Insofar as the old partial minimal demands of the masses clash with the destructive and degrading tendencies of decadent capitalism – and this occurs at each step – the Fourth International advances a system of *transitional demands*, the essence of which is contained in the fact that ever more openly and decisively they will be directed against the very foundations of the [capitalist] regime.' This is because, for a secure future, the key aspects of the 'old' need to be completely replaced by new ways of being – for instance, by ending the private ownership of the most important economic drivers. Essentially, to defend reforms, we need to go *beyond* reforms and reformism, towards an ecosocialist 'Grand Transformation' that creates a truly socially-just and ecologically-sustainable world. The encouraging news is that the number of those realizing that such a change is desperately needed is growing by the day: 'For many years the word anti-capitalism was part of the vocabulary of small revolutionary organisations. Now it is part of the language of millions.'[7]

And there's more!

Borrowing the best-known catchphrase of Irish comedian Jimmy Cricket, this section presents just some of the more recent evidence showing why a revolution is now urgently needed – in case some readers need still more reasons to join the struggle to end capitalism! This recent evidence – of the increasingly fatal irrationality of capitalism – is strong confirmation that we are indeed now in the midst of an existential crisis, as a *direct* result of the operations of the capitalist system. For instance, original estimates of the damage caused by Hurricane Milton in October 2024 were put at around $50bn. Then, following the hurricane moving eastwards, a storm surge hit the coast of Florida. Climate experts did not hesitate to explain the ferocity of this hurricane by referring to the high temperatures of both the ocean and the air in the Gulf of Mexico. Such exceptional warming was down to the continued burning of fossil fuels, which makes such extreme weather events more likely – and more extreme.[8]

What makes such extreme weather events a clear sign of the irrationality of the capitalist system is that forms of renewable energy are now *much* cheaper than fossil fuels – but capitalism is making too much money out of fossil fuels, and has too much money invested in that destructive industry, to accept a move to renewables. One of those to point out how far renewable energy – especially solar energy – has progressed as regards cost is Alan Thornett, who sees solar power as 'the key to the future of the planet.' The cost of solar panels has dropped by 90 per cent in the last decade, whilst new technological developments have now made it 'the most versatile and easy to maintain of all the technologies available.' According to Thornett and others, 'We are witnessing what is increasingly referred to as the Dawn of the Solar Age.' Michael Löwy is another to stress the importance of solar energy: 'Solar energy, which has never aroused much interest in capitalist societies (not being 'profitable' or 'competitive'), must become the object of internsive research and development and play a key role in the building of an alternative energy system.' This picture is confirmed by Malm and Carton, who point out that the price of renewable (or 'flow') energy – especially for solar – had been in something of a 'free fall' when compared to the price of fossil fuel (or 'stock') energy. So much so that 'the prices of stock and flow assumed almost the shape of a scissor: one blade going up, one down.' More importantly, the drop in renewable energy was much sharper than the rise in fossil fuel prices. Yet, despite such developments, the extraction and burning of fossil fuels continues to rise.[9]

This capitalist reluctance to invest to clean emergy is because there is no real 'value' in renewable energy compared to fossil fuels – essentially because there is very little 'social labour' involved in accessing renewable energy. The 'logic' of capitalism is simple: where there is no 'value', there is no profit. As Malm and Carton point out: 'No one has yet made a penny from producing sunlight or wind.' That is why, despite countless examples of increasingly-frequent and ever-more intense extreme weather events, and despite what are now yearly COP meetings, there has been no real curtailment of the burning of fossil fuels – and it's why the fossil fuel corporations are still ploughing large parts of their record profits into yet more fossil fuel projects. These amount to massive assets – assets which capitalism as a whole cannot afford to see 'stranded'. Yet, as far as people and planet are concerned, we cannot afford for those assets *not* to be 'stranded.' Marx's use of terms like 'monster' to describe capitalism – and terms like 'vampire' and 'vulture' – could not have been more appropriate. What capitalism *is* prepared to accept is 'overshoot' – a scenario in which humanity breaches 1.5°C, and 2°C or even 3°C, as long as fossil fuel assets aren't stranded. Thus, in reality, governments are not serious about 'mitigation' – i.e., coming off fossil fuels. Instead, they buy into capitalism's preferred option: 'adaptation, carbon dioxide removal and geoengineering, alone or in mitigation.' As Malm and Carton conclude: 'sooner or later, one way or another', humans will have to recognize that 'the political destruction of fossil capital' is vital – the alternative to doing so is 'unlimited combustion, [and] the maximum realisation of the potentials for catastrophe.'[10]

In fact, the deadly impact of capitalism's preferred trajectory can now even be calculated in human lives, after the publication of a study in July 2021, which showed that one million tonnes of CO_2 emitted in 2020 would result in 226 deaths. However, the author of the study, R. Daniel Bressler of Columbia University, pointed out that his figures were 'serious underestimates, since he only factored in mortality from a single climate hazard.' Malm and Carton used Bressler's figure to calculate that the oil extracted and burned as a result of the EACOP operation – owned by Total and other companies – would 'each year cause the death of 7,661 human beings.' Bearing in mind that Bressler's estimates were on the conservative side, Malm and Carton conclude that '7,661 people killed per year would be an unrealistic minimum.' Which is why they argue that, in today's worsening climate crisis, taking more fossil fuels out of the ground is an act of violence, firing 'carbon bombs… indiscriminately into humanity' – especially that part of humanity living in the global South.[11]

That reality of fossil fuel capitalism in particular – and of capitalism in general – is surely proof enough that its 'insatiable quest for profits' and its productivist 'logic' is 'leading us into an ecological disaster of incalculable proportions.' Furthermore, it is inflicting an increasing number of harms and calamities on humans, as well as on the other inhabitants of planet Earth. In essence, 'the dynamic of infinite "growth" brought about by capitalist expansion is threatening the natural foundations of human life on the planet.' All the evidence points to there needing to be 'a reversal, not merely a slowing down, of the underlying trends of the last few centuries.' As Sweezy and others have argued, such 'in a nutshell, is the meaning of revolutionary change today.' Whilst some measures of reform – 'no matter how desirable in themselves – can, at best, 'slow down the fatal process of decline and fall', that is all they can do. What we have to do is: 'replace capitalism with a social order based on an economy devoted not to maximizing private profit and accumulating ever more capital but rather to meeting real human needs and restoring the environment to a sustainably healthy condition.'[12]

The real harms that a small greedy minority, and their capitalist system, are doing to millions of people and to the planet has been abundantly clear for some considerable time. A whole swathe of crises – climate, ecological, economic, social, and political – are continuing, overall, to go from bad to worse. However, we also know how those crises have been created, and how and why they are still being perpetuated – as the saying goes: knowledge is power! Ultimately, what we need to do is turn that knowledge – and the justifiable anger – into revolutionary action. In the end, what we really *do* need is 'System Change' – *not* 'System Reform.' Because, as long as the capitalist system remains broadly the same, it will always try to return us to the 'muck of ages', as Marx and Engels once described capitalism's destructive legacy. And, as Steinberger has said: 'The hour is late.' What's required to turn things around is a movement strong enough to bring about a revolution – such an event is not only necessary because nothing less can bring about the changes needed, but also because '[those] *overthrowing* it can only in a revolution succeed in ridding [themselves] of all the muck of ages and become fitted to found society anew.' In 1938, Trotsky spoke of the urgent need for a socialist revolution, if the mounting horrors of capitalism were to be ended: 'without a socialist revolution, *in the next historical period at that*, a catastrophe threatens the whole culture of mankind.' But what seems obvious, as the last pages of this book were being written, is that the revolution we urgently need now is an *eco*-socialist one. So, how do we get there?[13]

Building the movement

Yet, even today, despite all the mounting evidence of how destructive – at its core – capitalism is, many activists in climate and social movements, and in centre-left parties, are failing to truly grasp just what is likely to be coming round the corner if serious, *revolutionary*, action is not taken, on multiple crisis fronts, in the next few years. To put it, bluntly, the full-spectrum emergency we are facing requires full-spectrum resistance. To build the kind of movement that is strong enough to resist, and then bring about what Brand and Wissen call the 'great social-ecological transformation', will need to be broad, pluralist, tolerant, democratic, clearly-focussed, and determined – and one that can, amongst other aspects, encompass anarchism's 'libertaire "enthusiasm" and Marxist "soberness"', as well as all those who align with neither the black nor the red flag of revolution. It will also need to be united – not 'united' in the sense of being uniform, but 'united' around achieving common goals. One crucial requirement in building that unity will be 'connecting supposedly different struggles with one another' – not just those concerned with the climate and ecological crises, or social inequalities, but also those challenging structural sexism, racism, homophobia and transphobia as well as. One idea of what such a movement should look like is referenced by Michael Löwy, who sees the views of Jorge Riechmann on how to construct a winning coalition – or 'gathering of the tribes' – as being particularly apt: 'This project cannot reject any of the colors of the rainbow – neither the red of the anti-capitalist and egalitarian labor movement, nor the violet of the struggles for women's liberation, nor the white of non-violent movements for peace, nor the anti-authoritarian black of the libertarians and anarchists, and even less the green of the struggle for a just and free humanity on a habitable planet.' Also essential will be involving LGBTQ+, anti-racist and anti-fascist movements. According to Brand and Wissen, that would allow us to move as quickly as possible towards 'a solidary mode of living that is aligned with the goal of a good life for all people on our planet', instead of the separateness and alienation currently experienced by so many as a result of capitalism's dominant 'imperial mode of living'. The good news is that history has provided examples of how it is possible to build a world other than the one constructed and imposed by global capitalism, which 'makes the people pay for its crises' caused by its 'destructive uncontrollability', and which has now turned even democracy itself into a 'commodity' to be bought and sold.[14]

As Naomi Klein has said: 'only mass social movements can save us now' – movements that will be strong enough 'to block the road, and simultaneously clear some alternate pathways to destinations that are safer.' Crucial to success in clearing such 'alternate pathways' will be constructing a movement – or movements – that knows what it wants to achieve. Because, as Michael Lebowitz has said: 'If you don't know where you want to go, no road will take you there.' Furthermore, as Daniel Singer warned in 1993: 'The absence of a vision for the future, ... of a radical alternative, cripples and paralyzes the potential movement from below and may divert it in dangerous directions.' We clearly need to get on a road that takes us, as quickly as possible, to the 'great transformation' envisaged by Brand and Wissen. Because the longer capitalism's destructive greed and short-termism hold sway over planet Earth, the more destruction it will continue to unleash, and that will make it harder to bring about the necessary changes. However, as Lebowitz also said: 'knowing where you want to go is only the first part; it's not at all the same as knowing how to get there.' Getting there – effecting a transformational break with the capitalist system – will not be easy, and will require the coming together and co-operation of all 'emancipatory social movements.' Only that will allow the birth of 'a new civilization that is more humane and respectful of nature.' Arguably, if any of the goals of such a movement is more important than the others, it is the latter of the two just mentioned. For if the natural world is destroyed beyond a certain level, the entire nature of human existence – regardless of attempts to achieve social and economic equality – will be at risk. It is now a matter of extreme urgency to restore a more reciprocal and harmonious relationship, or metabolism, between humans and the rest of the natural world. However, it is also increasingly obvious that any movement(s) attempting to do so will be mounting a revolutionary challenge to capitalism. Yet developing a movement that seeks such a change should *not* be impossible. Even if the word 'ecosocialism' is still relatively new to a lot of people, many aspects of what is now called ecosocialism aren't in reality that new – although many of its strands were even either unknown or ignored by many socialists in the previous century. As many have convincingly argued since 2000, revolutionary ecosocialism is in fact 'traceable to the long struggle for socialism and ecology beginning in the nineteenth century.'[15]

One of the first things such a revolutionary movement will need to do is to dispose of the cynical and pessimistic 'argument' that people, because of 'human nature', are inherently selfish and greedy, thus making any attempt to create a better world doomed to inevitable failure. It should come as no surprise that one section of the population in particular likes to propagate this myth: the one per cent whose greed is driving this planet to the edge of climate and ecological breakdown! Humanity as a species would not have survived – let alone spread across the Earth – if we had been nothing but selfish and greedy. Whether as hunter-gatherers – or later, as early subsistence farmers – caring, sharing and co-operation have been a vital part of our history. As Neil Faulkner pointed out: 'Were human society governed by greed and selfishness, a war of all against all, we would never have left the Stone Age.' So how do we use this knowledge to get enough people to reject the prevailing negative and capitalist ideology on human 'nature', which is so incapacitating and demobilising? Part of the answer to doing that will involve creating a movement in which people are encouraged to be 'outgoing, proactive, socially engaged, and politically committed' – in marked contrast to existence under capitalism, which for many is often 'alienated, privatised, [and] narcissistic.' A movement which operates in such a way would already be creating – in embryo – 'a new way of being' in readiness for life after ushering in a great post-capitalist transformation. It will also give members of that movement the sense of solidarity and strength needed to persist in what already is a difficult struggle – and which will become even more so.[16]

'Scared of us'

Any movement that is going to be able to bring about a successful revolutionary transformation will clearly also need to be strong enough, in the words of Thee Faction – a socialist RnB rock group – to make the one per cent 'scared of us.' Because, as they point out in their song 'Scared of Us', even as regards trying to win reforms:

We don't get nothing
Till they're scared of us!

Of course, the big question is how to get the one per cent sufficiently 'scared of us' that they come to realize that we *do* have the power and determination to achieve a world based on sustainability, equality, justice, democracy and peace? One proven way to is to struggle: history is littered with examples in which struggle has resulted in meaningful reforms, if not outright revolution. Crucial to building such successful struggles is undoubtedly bringing different groups and sections of society – currently depressed, exploited and oppressed by capitalism – together to unite in struggle for a better world than the one capitalism is imposing on us. The oppressed uniting in struggle for a common cause, however, does more than just make the one per cent 'scared of us' – as noted by Faulkner above, the comradeship and strength gained in collective action actually helps people overcome their isolation and alienation, builds confidence and consciousness, and strengthens the determination to win the struggle.[17]

From Marx onwards, associated working people – the organized working class – have been seen as *the* agents for bringing about a revolution that will allow people to move beyond the destructiveness, inequalities and constraints of capitalism to a better post-capitalist world, whether that be socialism or communism. However, the reality now is that, while the organized labour movement will still be significant players in bringing about the revolutionary and ecological transformations needed to move beyond capitalism, there are now other 'agents' which are also important – both in developed centres of the global North, and in the periphery of global South. Nonetheless, the 'working class' is still very much with us – despite those who have argued since the 1980s that it has all but disappeared. But today that class comprises 'those of us who rely on wages to live, who own few assets or no assets, those that carry out unpaid essential work in the home and in care, those that are exploited for money by others.' Thus defined, this working class is clearly 'the largest group of people on the planet' – and will thus have a crucial role to play in achieving the ecosocialist revolution we need. That part of the working class organized within trade unions – including those in 'middle class' jobs such as teaching and the medical pofessions – will form a particularly significant part within such a broad movement; and the policy stances they take will be crucial to a successful struggle against global heating and the climate crisis. For instance, the

Campaign Against Climate Change (CACC), formed in 2001 with an important Trade Union Group as part of it, has been a positive development. Encouragingly, it is backed by six national unions, and its *One Million Climate Jobs* pamphlet and campaign has gained much support from the TUC, because of the way it sets out how a just transition away from fossil fuels is possible via the creation of good, well-paid, 'green' jobs.[18]

However, while Matt Huber has correctly pointed out that 'climate change is a class issue', it is important that those on the more traditional left should avoid dismissing the significance of various other social movements – especially the growing climate movements. Climate and social movements such as Extinction Rebellion, Insulate Britain, and Just Stop Oil (JSO) — along with campaigning groups like Greenpeace and Friends of the Earth (FoE) – are also hugely important. Most recently, in March 2024, JSO officially issued an explicit call for revolution with the launch of their new organization – consisting of four sub-groups – called 'Umbrella', to take action 'across a range of issues, including but not limited to the climate crisis.' The preamble to the launch had this to say: 'Our world is rapidly heating beyond safe temperatures. The collapse of everything we love is becoming inevitable and, for millions of people around the world, it's already happening…The situation demands that we replace our entire broken political and economic system. We cannot continue to only demand one small policy change. *We need a revolution.*' Just Stop Oil, while continuing to struggle against all new oil, gas and coal projects, sees Umbrella as becoming the 'home of a revolutionary community.' The following month, April 2024, saw one of those four sub-groups – Youth Demand – take its first action. Chiara Sarti, a spokesperson for Youth Demand, stated that 'because the current political system is broken beyond repair', revolutions are to be expected – so the crucial question to be posed was 'are we going to get a fascist type of revolution or are we going to get something better, get a democratic revolution based on nonviolence?'[19]

Undoubtedly, the courage and sacrifice of so many climate activists, who've been willing to be arrested and imprisoned, have massively raised general awareness of the climate and ecological crises. A successful revolutionary struggle will thus need to recognize the central importance of such groups – and seek ways to combine them with organised movements of workers in the struggle ahead. Thus, it is encouraging that Extinction Rebellion UK now tends to define the 'System' more precisely as a 'Commercial Economy' that puts profits before people and the planet. Also, XRUK has re-launched a trade union wing – Extinction Rebellion Trade Unionists (XRTU) – which, along with JSO, has been heavily involved in supporting workers' struggles, drawing out the ecological dimensions of disputes and promoting ideas of a just transition. Only such powerful social movements, in alliance with the labour movement, will be able to pull what Walter Benjamin called the 'emergency brake' on the capitalist train that is ever-rapidly taking us down the rails to catastrophic climate breakdown and social barbarism

Internationalism

However, given the scale of the multiple crises the whole world is facing, and that capitalism is a global system, it is clear that as well as taking action within Britain, it is also vital to make international connections and to co-operate with like-minded organizations across the world. Ecosocialists thus attempt to unite people across borders, to build a global movement that is able to develop practical solutions at all levels: local, regional, national, continental and global. Part of this will involve embracing 'the indigenous peoples around the world along with major social movements, such as La Via Campesina and the Brazilian landless Workers Movement (MST).' Ultimately, if we are to mitigate and then end today's multiple crises across the globe, we need an international ecosocialist coalition of radicals – rather like the Zimmerwald Movement which developed in 1915-16 during the crisis of WW1. The formation of the Global Ecosocialist Network in 2020 – which has as its slogan the same one launched by XRUK two years previously: 'System Change, Not Climate Change' – in order to raise awareness of ecosocialist arguments, was a really useful step in the right direction.[20]

As part of that internationalism, and to make that fundamental global economic shift from capitalism to a collective and democratic ecosocialist society, there will need to be a meaningful recognition of the historical and contemporary debts that the North owes the South – including a commitment to unconditional transfers of wealth and technology from the global North to the global South. For instance, cancelling all debts 'owed' by countries in the global South. This will be crucial in achieving climate justice across the world; in effecting just transitions to systems based on satisfying people's real social needs; and in ending the wasteful and destructive production-methods of capitalism – thus establishing human life on a truly ecologically-sustainable basis. As Mikaela Loach, and many others, has argued, because of decades, and even centuries, of colonialism and imperialism, 'many of the countries most impacted by climate change do not have the ability to be resilient'; while the 'wealth inequality' which has resulted from such exploitation means that capitalist corporations are currently able 'to pollute as much as they do' in poorer countries. As a result, 'sacrifice zones' are being created, with 'entire areas deliberately destroyed' in order to facilitate capitalist profits being obtained in environmentally-destructive ways no longer permitted in Europe itself. It is such imbalances of power which are resulting in the exacerbation of the climate and ecological crises. Though, as pointed out by Kate Soper, 'the ecological "debt" incurred through ongoing unequal exchange' cannot be understood in purely monetary terms. For instance, the compensation for the damage done to Africa as a result of Europe's earlier mass enslavement of its people is incalculable. Nonetheless, solidarity with current struggles in the global South is essential. In the global South, contemporary capitalism continues to use the same brutal 'blood and fire' methods which it used to such ruthless effect during its birth in Europe. Communities trying to resist the impacts of capitalism's globalized fossil fuel economy today are thus still facing capitalism's 'harshest means: murder, militarisation, land grabs, displacement and ecological destruction which causes impoverishment and forces local communities into working cheaply for the industry.' In Latin America – in countries such as Ecuador and Bolivia – there is 'intense and widespread resistance to neo-extractivist pressures.' Often this resistance is being led by Indigenous groups, and is sometimes closely connected 'to a history of resistance to colonialism and neo-imperialism.' It is thus vital for global anti-capitalist movements to establish meaningful and practical solidarity with such examples of popular resistance wherever they occur.[21]

Not surprisingly, the worsening climate and ecological crises are resulting in an increasing number of people fleeing the impacts of global boiling and its associated extreme weather events – as well as from poverty and war. Thus, internationalism also requires movements to show solidarity with such refugees and migrants – including resisting attempts by far-right populist and outright fascist groups to stir up hatred of such victims of capitalism. As is increasingly clear to many, with corporations, governments and mainstream parties refusing to put climate protection before capitalist profits, the choice facing us – and most of the other species on this heating-up planet – is, quite simply: 'either ecosocialism or capitalist barbarism and extinction!' Thus, whilst it's necessary to fight hard for all the reforms, mitigations and policies we can force from our respective national governments, we will ultimately have to make a decisive break with the logic of capitalism itself on a global scale. As Marx would (probably!) have said if he were alive today:

> *'People of the world unite, rise up, and ACT!*
> *You have a planet to save!'*

Time to act!

Quite simply, in order to get the revolutionary 'Great Transition' we need – from the in-built destructiveness of capitalism to an ecologically-sustainable and socially-just society – we do indeed have to act! In doing so, we will undoubtedly take inspiration from the revolutionary and ecological ideas put forward by a range of radical environmental and revolutionary thinkers – including people such as Epicurus, through Karl Marx and Friedrich Engels, William Morris, to more recent writers such as Rachel Carson, Elizabeth Kolbert and so many more. Such a revolutionary transformation may – at this point in time – seem improbable. However, one thing that should be clear by now is that capitalism – in *any* form – cannot be allowed to continue for much longer 'if human civilization and the web of life as we know it are to be sustained.' Such a position includes recognizing false hopes put forward by capitalism's defenders – such as the suggestion that capitalism will eventually 'dematerialize' its production by 'producing' services rather than commodities, thus reducing 'its reliance on matter-energy throughput'; by producing wastes that the natural world can better tolerate; and by better recycling. Yet, as Paul Burkett, and others, have shown, such claims assume that production of services won't require a 'growing material base.' As many have already discovered, the greater amount of information and entertainment now available is dependent on increasing 'matter-energy throughput' via computers, scanners, mobile phones and media players of various kinds – not to mention the many components and materials used in the manufacture of such products, or the 'rapid rates of obsolescence' and thus increasing waste-materials dumped upon the natural world.[22]

As has been observed by many, regarding the multiple crises generated by capitalism, we are 'close to midnight…. We have no time to lose.' There are no guarantees that we will succeed – but one thing is certain: if we don't try to build a better world by making a 'great transition', things are not likely to change for the better. In fact, it's highly likely that things will be even worse than the period before and during the Second World War. This is why it is so important to get active in relevant campaigns and struggles – now – in order 'to create a different future', and so 'become a maker of history not an object of history.' One of the best – although *not* the only – way to do so is also to join a revolutionary ecosocialist organization that is fully committed to establishing an ecologically-sustainable economy, which puts planet Earth and people before profit, which ensures our collective resources are put at the service of all, and which attempts to do so via mass participatory democracy, international solidarity and co-operation. However, 'talk changes nothing' – the truly important thing is to start getting active. For, as Neil Faulkner pointed out: 'First is the deed. It is necessary to act, for only by acting can we change reality.' Or, as Marx wrote: 'The philosophers have only *interpreted* the world, in various ways; the point is to *change* it.' Marx also noted that people and their consciousness change in the act of changing the world. Thus, in the struggle for the ecological red-green revolution we need, in the struggle against injustice, and for a radically transformed world, we will connect with, and learn from, others. We will also deepen our understanding of the roles of the state, the police, the courts and the media in a capitalist society – as many strikers have learnt over the decades, and as many Just Stop Oil activists are learning now.[23]

As Michael Löwy, a leading ecosocialist thinker, has said: 'preserving the ecological equilibrium of the planet and therefore an environment favourable to living species, including ours, is incompatible with the expansive and destructive logic of the capitalist system.' Quite simply, if not ended, capitalism could well collapse the planet into catastrophe. To prevent that, we need a revolutionary transformation. As a first step towards that transformation, we need to abandon the propaganda which tells us that capitalism is the 'end of history', and that what exists now is the only possible way of living. As if to prove the fallacy of such an 'argument', in May 2024, *The Guardian* published the findings of the USA's National Bureau of Economic Research (NBER), which echoed those of an earlier study: 'The economic damage wrought by climate change is six times worse than previously thought, with global heating set to shrink wealth at a rate consistent with the level of financial losses of a continuing permanent war.' Marx and Engels in *The Communist Manifesto* said the working class were the 'gravediggers of capitalism' – they, and the climate movements, need to undertake (pun intended!) that role now: otherwise, capitalism will be the 'gravedigger' of huge swathes of life on Earth. Having seen how capitalism came into being, what destruction and cruelty it has already inflicted on people and planet in its relatively short existence, and the many more recent unsustainable harms it continues to impose – in its never-ending pursuit of ever-greater profits – may well lead some to think that, no matter how 'evil' capitalism is, any major transformational 'System Change' seems impossible. Yet Nick Estes – amongst many others – is absolutely right to state that such a revolutionary change is vitally imperative: 'We demand the emancipation of the earth from capital. For the earth to live, capitalism must die.'[24]

Such a judgement has been echoed by so many; and an increasing number are concluding that, in the end, the only viable alternative on the horizon right now – 'to [capitalism's] current process of destruction of the natural foundations of life on the planet' – is ecosocialism, based on Marx and Engels's earlier critique of how capitalism works: 'of the blind logic of value, of the brutal subjugation of human beings and nature to the imperatives of capital accumulation.' Such a judgement – on the incompatibility of capitalism's inherent productivism with an ecologically-sustainable way of living on Earth – is one shared by Daniel Tanuro: 'The system's bulimic consumption of energy derives from its logic of unlimited accumulation, which is the main cause of the ecological crisis and the climate crisis'.

Another to draw the same conclusion is Rebecca Solnit: 'nothing less than systemic change will save us.' As hard as the struggle seems at times, we must keep trying. To reference what the anti-fascist resistance leader Victor Laszlo says to Rick Blaine in that classic Second World War film *Casablanca*: 'If we stop fighting…, the world will die.' That same sentence and sentiment applies with at least equal force to today's climate and ecological crises. We must refuse to give in and give up: there's too much at stake – especially for younger and future generations.[25]

Even just asking for an ecologically-sustainable world – never mind one that is also socially-just – increasingly challenges the one per cent: their corporations, their obscene levels of wealth, their property rights and their political power. But, for the Earth to live, we have not only to maintain those challenges, but must push them to the point where we can indeed effect a revolutionary 'System Change' or 'Great Transformation'. Of course, those at the top of any system always argue that real change is neither desirable nor possible – but then they would, wouldn't they?! For those who despair of ever being able to bring such a transformation about, it is worth recalling that revolutions can sometimes come about rapidly – and peacefully – when it seems that the possibilities for such changes seem remote. Take, for example, the 'velvet revolutions' in 1988-89 that, in just a few years, swept aside the one-party states in Eastern Europe. The bottom line is simple: if we don't continue to struggle, we – and all the other Earthlings who live on this planet – will lose.

Drawing up 'any detailed blueprint' for the exact unfolding and operation of the 'global-epochal transition' (i.e., revolution) we and the Earth need would be impossible. For a start, there will undoubtedly be setbacks and problems – as well as successes – on the road to the creation of a 'communal system system of sustainable human development.' Nonetheless, it's becoming increasingly clear that only through a 'real communality, in which people gain control over the social conditions of their existence' – as opposed to being under the control of capitalism's exploitative and profit-oriented system – will human society be able 'to regulate its metabolic interchange with nature in a healthy and sustainable way.' An essential part of that struggle to create such a society will be the question of democracy and freedom. Given that many of the crises we face are national or global rather than just local, there will need to be a mixture of direct and indirect democratic systems – with decision-making at the appropriate level, whether 'local, regional, national, continental [or] global.' Under such a democracy, universal suffrage, 'a multiparty system, freedom of the press, and oversight and recallability of the elected' will be essential components, in order 'to entrust as many powers as possible to the base and foster local initiatives.' It is impossible to know what forms politics will take in the future but, to avoid past mistakes, freedom will have to be at the centre of an ecosocialist society. As the surrealist poet and libertaire Marxist Benjamin Péret wrote, freedom constitutes 'for the mind as for the heart, the oxygen without which it cannot survive. If the physical being cannot live without air, the emotional being can only wither and deteriorate without freedom.' This is why, for carrying through a revolutionary transformation which has a strong 'human face', self-organization and anti-bureaucratic structures are – and will continue to be – of paramount importance. As part of that, as Löwy and Besancenot have argued, it will be essential to '"re-individalize" the communist project [and] "collectivize" anarchist ideas', in order to follow 'a revolutionary humanist path' to a better world.[26]

The crucial point to take away is that the goal – of collectively creating a society that 'produces less and differently, transports less, cares more for people and nature, shares wealth and decides together' – *is* achievable Furthermore, the greatest obstacles to the creation of such an ecologically-sustainable *and* socially-just world – in other words, an ecosocialist world – are *not* the power of the capitalist system, or the wealth of the capitalist class, or the machinations of capitalism's political and intellectual supporters. The biggest obstacle 'is the conviction, held by so many millions of people, that change is impossible.' The good news is that 'ordinary' people – who *believe* in the possibility of change – really do have the power to make revolutions. Many of us have now become increasingly indignant at the capitalism's multiple injustices, and many have engaged in acts of resistance. Though such actions often begin as *individual* acts of defiance, with the first refusal to disperse or unlock being just one example of personal strength, they have wider, and more *collective* significance. They are, in fact, 'a source of personal enrichment' which enables people to begin shaking off 'the chains of contemporary alienation.' This kind of liberty can then translate into people undertaking collective activism. As Grace Blakely has pointed out: 'The moment that we let go of our collective belief in the inevitability of the current system is the moment that we start to build a new one.'[27]

CONCLUSION: HOPE AND HISTORY

Your opponents would love you to believe that it's hopeless, that you have no power, that there's no reason to act, that you can't win. Hope is a gift you don't have to surrender, a power you don't have to throw away. And though hope can be an act of defiance, defiance isn't enough reason to hope. But there are good reasons.[1]
Rebecca Solnit

Just over fifty years ago, Barry Commoner – an early environmentalist – warned that 'the environmental crisis reveals serious incompatibilities between the private enterprise system and the ecological base on which it depends.' As 2024, and this book, draws to a close, it is surely appropriate to recall what Ian Angus said about Commoner's early alarm call: 'it is time – it is *past* time – to hear his warning and change that system.' Capitalism is increasingly turning the present and the immediate future into a disaster-zone, and the very seriousness of today's crises should compel us 'to press ahead to something better.' The good news is that history shows that, sometimes, things can and *do* change for the better – and it also shows that people 'inspired by hope for a better world, can achieve incredible things.' That's an important fact to hold on to in these dark times.[2]

States of denial

There are those within both the environmental and the ecosocialist movements who, because *they* have heard the warnings uttered by Commoner and others, wonder why many more people haven't also heard – and responded – to those warnings. The sad fact is that, as every week brings yet more evidence of the harms being done to the planet, much larger numbers of people *are* becoming aware of the destructive actions resulting from the insatiable greed of extractivist and productivist capitalism. The problem isn't so much that people are unaware, it's that the *scale* of what's being done, and thus of what's required to heal the planet's fragile ecosystems, is so overwhelming. Consequently, 'they get very sad, they get very quiet.

So quiet that protection of the environment … and [imagining] a future for their children doesn't even make it onto a list of their top ten concerns.'[3]

As Stan Cohen wrote back in 2001, when people are presented with information that's really too disturbing or too threatening to be fully absorbed or acknowledged, the information is somehow 'repressed, disavowed, pushed aside or reinterpreted. Or else the information "registers" well enough, but its implications … are evaded, neutralized or rationalized away.' In other words, it's often an unconscious defence mechanism to protect ourselves from unwelcome or disturbing truths. Because once we *acknowledge* that we truly 'know' something, we have to make a decision: either to do something about it – or to refuse to act, despite knowing. Doing the latter often results from too much information about doomsday warnings – whether about destruction of rainforests, melting glaciers, the mass extinction of species, increasingly gross inequalities of wealth, or the spread of far-right and fascist ideas. All that can then lead to grief and then the kind of despair that leads to inaction.[4]

Thus, in effect: 'we manufacture [oblivion] for ourselves to keep us from looking environmental problems straight in the eye.' Suppressing our natural responses to disasters and crises can, in fact, be seen as an aspect of 'the disease of our time' or a 'dangerous splitting', resulting from a refusal to fully acknowledge the truth. This split then 'divorces our mental calculations from our intuitive, emotional and biological embeddedness in the matrix of life.… [and] allows us passively to acquiesce in the preparations for our own demise.'[5]

Defeating despair

There are so many scary examples of the increasingly-dangerous environmental harms and ecological destructions that capitalism is inflicting on the natural world that it's not at all surprising that many feel genuine – and mounting – grief, fear and despair. Yet, both Trotsky and Gramsci, despite also facing very difficult and challenging times, counselled *against* despair. According to the latter, those most likely to avoid despair were those who were 'deeply convinced' that the sources of their 'own moral forces' were in themselves: that would ensure they never fell 'into those, vulgar, banal moods, pessimism and optimism.' As Gramsci explained in a letter to his brother in 1929, his own state of mind both synthesized and transcended those two feelings: 'my mind is pessimistic, but my will is optimistic.' As he never entertained unrealistic illusions, he was seldom disappointed; instead, he was 'armed with unlimited patience – not a passive, inert kind, but a patience allied with perseverance.' Thus, if we are to put things right, we need to move beyond our justifiable tears and fears. In particular, we need to move to *feeling*: feeling love for planet Earth and for all the Earthlings that live on it – *and* feeling a positive anger against those 'systems' that are trashing planet and people. That's how we'll start to push back against the one per cent – and then move on to restoring and nurturing the natural world so that we can re-build an ecologically-sustainable world, and also create socially-just societies. However, as well as having dreams and hope being crucial, we also need to be realistic: there are no guarantees that an ecosocialist revolution will come about. As Angus said: 'It will only happen if people consciously decide it is necessary, and take the steps needed to bring it about.' At the beginning of *The Communist Manifesto*, Marx and Engels warned that struggles for new ways of living – provoked by system crises – always end 'in a revolutionary reconstitution of society at large, *or* in the common ruin of the contending classes.'[6]

What we do know is that, as Gramsci warned, we must not give in to despair and give up struggling. Because although we *may* lose, it is now absolutely certain that if we don't continue to struggle, we – and the planet – *will* most definitely lose. As Gramsci said: 'It is necessary, with bold spirit and in good conscience, to save civilization. We must halt the dissolution which corrodes and corrupts the roots of human society. The bare and barren tree can be made green again. Are we not ready?' It is precisely Gramsci's approach – which shows the extent to which he was able to transcend 'the dichotomy between pessimism of the intellect and optimism of the will' – that's needed today. This is not an argument in support of a 'facile optimism' based on unrealistic grounds, but an argument for what Leo Panitch, in his 2017 article 'On Revolutionary Optimism of the Intellect', has called an 'optimism of the intellect' that 'involves bringing reason, ethics, imagination to bear on how to realize optimism of the will.' It is such realistic optimism – 'intellectually tempered by a sober recognition of the great barriers to positive transformative change' – that gives us the strength to go on making 'whatever positive contribution we can to overcoming those barriers, including in ourselves and our institutions.'[7]

Such 'optimism of the intellect' will allow us to go on to develop an 'optimism of the will' to ensure we can – and will – continue to struggle for a better world. As an example of such necessary 'optimism of the intellect', Panitch quotes from Thomas Dewey's 1916 essay on 'Progress', written in the midst of the horrors of the First World War – when, as Dewey observed: 'Never was pessimism easier.' The important point made by Dewey was that improvements and 'progress' do not come automatically, but depend on 'human intent and aim and upon acceptance of responsibility' for bringing them about. In 1916, Dewey concluded that what stood 'most in the way of progress were not the forces of conservatism and reaction but rather the much more common disbelief in the possibility' of change and improvement. According to Panitch (and many others, including Che Guevara), it is such 'common disbelief' which is still 'the greatest barrier that optimism of the intellect faces' – and thus the greatest barrier to developing and embracing the 'optimism of the will' we also need to bring about the 'Great Transformations' so urgently required by people and planet. Despite past failures, and despite ever-increasing warnings about the approach of climate and ecological disaster, there remain grounds for hope that, via 'collective human agency', we *can* achieve the changes needed – along with evidence that there *is* still time to achieve at least some of our goals. The bottom line, of course, is that if we don't continue struggling for change – even if some of the goals appear improbable – things are very unlikely to change in the ways needed.[8]

Revolutionary hope

While it is undoubtedly true that the real potential for truly ecologically-sustainable and socially-just human societies is being betrayed by the lies and practices of capitalism and its supporters, having that knowledge, understanding and justifiable anger – on their own – will not be enough: we also need hope. As stated in the Introduction, this book is in essence based on a *realistic* hope of the possibility of achieving fundamental political, economic and social changes – despite the many deep and existential crises we're currently facing. The crucial thing about *realistic* hope is that, while it does not guarantee victory in the future, it can most certainly act as a detonator to release the energy we require to undertake the actions needed today. The quotation from Rosa Luxemburg – a great and independent Marxist thinker and activist

– that headed the previous chapter, shows that she too believed that, in the end, people and planet could be emancipated from the rule of capitalism. Significantly, she was able to maintain such hope even as the world was plunged into the horrors of the First World War.

A more recent revolutionary, who also retained such revolutionary hope in the possibility of a truly emancipatory revolution, was Che Guevara, who believed that 'the most beautiful quality a revolutionary can have' is to feel – deep within their being – 'all the injustices committed against anyone, anywhere in the world.' In the end, of course, we shouldn't oppose 'exploitation, oppression, massive violence against human beings, and massive injustice' just because we think that such struggles will be successful in creating a better and fairer world. Instead, in the words of Ernest Mandel, we should oppose such injustices simply because they are 'inhuman unworthy conditions. …[that] is a sufficient ground and motivation.' Or, as he said elsewhere: 'In the face of massive injustice, resistance and revolt – including individual resistance, but above all collective resistance and revolt – are not only a right but a duty.' Today – for our own sakes – those 'injustices' we need to feel must include those committed against *all* the inhabitants of planet Earth, not just the human species. Equally important, in view of the effort that will be needed to move on from winning reforms – and especially 'non-reformist' reforms – Guevara also stressed that temporary defeats were not important, because: 'What is decisive is the determination to struggle…, the awareness of the need for revolutionary change and the certainty of its possibility.'[9]

Such an approach is ultimately based on hope which, as mentioned elsewhere in this book, is a crucial revolutionary attribute. Many others have made a similar point about the importance of having hope that a better world *is* possible. Recently, Mikaela Loach – who has described hope as 'an active stance' – put it thus: 'We… have to completely and urgently reframe how we collectively think about this [combined] crisis, moving away from doom and gloom and into hope – a hope that can transform the world around us. This is what will not only mobilise the numbers we need to create change but also ensure that what is to come is a world centred on justice and liberation for all. A better world.' Such hope is not based on mere wishful thinking – as Loach has convincingly shown: 'the ground for revolutionary change has never been more fertile than it is now.' By now, the need for such a revolutionary transformation should be abundantly clear, as we have examined more than enough evidence to suggest – very strongly – that we are living in one of the darkest periods in Earth's history: not just as regards human existence, but also as regards the survival of millions of other species.[10]

However, there is more than one meaning of the word 'dark'. Yes, it can mean dangerous and depressing; but it can also convey the much less depressing sense of an 'unknown' future. This was the sense used by Virginia Woolf in her *Diary* for 1915, written at the height of the First World War: 'The future is dark, which is on the whole, the best thing the future can be, I think.' Yes, of course the future is 'dark' – but it's 'a darkness as much of the womb as the grave.' While, of course, it is possible that the future might turn out to be even worse than the present, it's equally possible that the future might well contain a much brighter, better and fairer world. And it's that very uncertainty and unpredictability that gives grounds for hope and optimism – and thus the impetus to *act*. If nothing else, surely the current situation strongly confirms what the World Congress of the Trotskyist Fourth International stated way back in 1995: 'there is no lack of reasons for keeping the flame of revolutionary hope burning.'[11]

While an increasing number of us can see just how bad things are, and can also see the need for a range of reforms and adaptations that could halt – and even reverse – some of the worst aspects, many of us are painfully aware that those reforms are still often blocked by the wealthy corporations whose short-term drive for increased profits is having such massive negative effects on people and planet. Thus, it can be difficult to look, with real hope, from this very 'dark' present into what is perforce a 'dark' future.

Kairos – the moment of transition

However, it is the very fact that the future has yet to be 'written' which should give us the necessary optimism to hope that we *can* change things for the better. Mention has been made earlier of Gramsci's comments about how, during his time, a new world was struggling to be born. In ancient Greek philosophy, the term *kairos* denotes a 'moment of transition', or right time, when a real transformation is possible. Every so often, history throws up such moments of transition, or turning points, when human agency – as a result of people acting collectively – can bring about a move to a new way of living. One with such realistic optimism about the future is Robin Kimmerer, a Professor of Environmental Biology. First of all, she lists some of the major problems currently facing us: 'famine for some and diseases of excess for others. The very earth that sustains us is being destroyed to fuel injustice. An economy that grants personhood to corporations but denies it to the more-than-human beings.' But, having done so, she asks the following questions – which many of us are asking today: 'What is the alternative? And how do we get there?' Her answer is an honest one: 'I don't know for certain'; but she sees at least one possible solution to creating a more sustainable and just world can be found in applying the more harmonious relationship with nature found in Indigenous societies. Which, at root, is precisely the essence of ecosocialism. However, one attitude that it's absolutely vital for us to *avoid* is the fatalistic belief that the future will inevitably be written by those powerful interests which are currently writing the present. In particular, we should take encouragement from the fact that history – including the history of the past fifty years – provides us with many examples when the future desired by powerful elites was *not* written in the ways they desired and felt confident would be achieved.[12]

The good news is that sometimes those futures desired by various seemingly all-powerful elites were instead written by 'ordinary' individuals, groups and movements who rejected both the present *and* the future those elites wanted – and who instead together constructed the future *they* wanted. One relatively recent example, mentioned previously, is that provided by the movements that toppled the 'communist' regimes of eastern Europe in 1989. The vast majority of those regimes, whilst increasingly unpopular for a variety of reasons, had seemed absolutely stable and entrenched even just a few years before.

While the mass demonstrations and protests, which hastened the end of the rule of those single-party states, were the highly visible – and seemingly spontaneous – expression of revolution, those events had been slowly developing as the result of private discussions, the production of subversive political and cultural writings, and long-time organizing. That largely unseen process was a classic example of how revolutions build – slowly at first, and then gathering increased momentum as largely-isolated individuals and small groups emerge from the underground to challenge the authorities.

This is precisely what Marx meant when he wrote how pre-revolutionary trends during periods of stagnation and even repression can suddenly erupt into open and *successful* revolution: 'our old friend…old mole [revolution], who knows so well how to work underground', can then suddenly appear.[13]

This is precisely how hope and history can – and do – intersect. When a situation is dire – like the one we're living through – memory becomes revolutionary once people remember the number of really significant changes, and especially victories, that there have already been. Sometimes, these have been systemic changes – sometimes they have been in the realm of cultural attitudes. Think, for example, of the successful struggles to end the traficking of enslaved people, the fight for votes for women, the struggle against apartheid in South Africa, and the winning of LGBTQ+ rights. Of course, we have to be realistic, and recognize that there have also been failures. The elites clearly prefer for the many to forget such successes and just remember the failures: doing the latter simply leads to acceptance that the 'real' world is *'immutable,* inevitable and invulnerable.' Thus, a lack of historical memory simply helps reinforce such a depressive view of life and our ability to change things for the better. Remembering history – the victories as well as the defeats – 'is where hope comes in.' This is one of the reasons why historians who refuse to forget are often some of the first victims of repressive regimes. I was fortunate enough to be a university student as the May 1968 events began to unfold in Paris – and which then triggered a global student revolt. One of the slogans which appeared on the walls of Paris was: 'Be realistic! Demand the impossible!' Today, demanding the end of capitalism is clearly 'realistic' – given how it's driving the climate and ecological crises, along with creating ever-greater economic and social inequalities. So maybe it's time to resurrect that slogan from 1968 – and the good news is that it's *not* an impossible goal to aim for![14]

Today, the contradictions of capitalism – especially its inability to address the climate and ecological crises – are much deeper and more apparent than they were in 1968. In fact, those contradictions are greater than they have ever been. In addition to the failure to address those existential crises, there are also recurrent slumps and recessions, increasing poverty, and nasty regional wars. There is also the increasing threat posed by far-right populism and fascism – which in large part arises out of the despair created by capitalism's failures and crises. All this is precisely why hope, despite these crises and emergencies, is so important: 'Inside the word *emergency* is *emerge*, from an emergency new things come forth.' This book is, to a large extent, based on the understanding that it is ecosocialism which gives the best grounds for such hope. This, in turn, is based on the historical fact that when the public – often referred to as 'the sleeping giant' – really wakes up to certain problems, it becomes a strong 'civil society' that can be, and *has* been, 'more powerful than regimes and armies.' As Rebecca Solnit has convincingly argued: 'Together we are very powerful, and we have a seldom-told, seldom-remembered history of victories and transformations that can give us confidence that, yes, we can change the world because we have many times before.' It is my genuine belief that we are now living in one of history's transitional periods – if I did not, I wouldn't have spent the best part of a year in putting this book together![15]

The *possible* dream

For some time, it's been clear to many radicals that 'a stark choice faces humanity: save the planet and ditch capitalism, or save capitalism and ditch the planet.' Despite this, one of the sayings that has been current within radical circles since the 1990s is that, for many people, 'it is easier to imagine the end of the world than to imagine the end of capitalism.' This is usually attributed to the Marxist theoretician Fredric Jameson, who first used it in 1994, and then again in 2003 – though Slavoj Žižek is also sometimes credited with the idea. Essentially, for many, environmental apocalypse seems much more likely than 'the triumph of a systematic economic alternative' to capitalism. As a result, the idea of history as something made by ordinary humans seems unimaginable under late capitalism, and so any real sense of a 'historical imagination is paralysed.' According to Mark Fisher, such 'capitalist realism' results in a widespread belief 'that not only is capitalism the only viable political and economic system, but also that it is now impossible even to imagine a coherent alternative to it.' The 'lesson' the one per cent would like us to draw is that, therefore, any popular resistance to capitalism is futile, and any attempt to build a better world is mere 'utopianism'. The way to respond to such negativity is to call to mind what Daniel Singer said in his 1999 book about the coming millenium: 'If any attempt to change society, and not just mend it, is branded angrily and contemptuously as utopian, then, turning the insult into a badge of hon-

or, we must proudly proclaim that we are all utopians.'[16]

As well as often painting even non-violent protests as both criminal and 'dangerous', the newspapers and TV stations – mostly owned by sections of the one per cent – also daily serve up the arrant nonsense that humans by 'nature', are greedy and selfish. But we don't have to abandon the dream of a better world, or digest such negative views of human potentialities – all we need to do is just call to mind some of the many examples of human empathy, compassion, generosity, and solidarity: not just towards other humans (including complete strangers), but also to other Earthlings. There are countless examples of people responding to major disasters in actively altruistic and community-spirited ways in the face of devastating earthquakes, floods, hurricanes – and even in war situations. Such situations also frequently throw up examples of resourcefulness, improvisation and creativity amongst survivors that give an alternative and *revolutionary* view of so-called 'human nature' – and this should give us the *realistic* hope that we *can* build a better future if we work together.

Sadly, most people live under a social and economic system that was created by, and continues to be based on, ruthless selfishness and greed: capitalism. The daily grind of merely surviving – and the constant emphasis on individual 'success' – tends to repress feelings of empathy and co-operation in many people, and, instead, to emphasize capitalism's 'morality' of greed and selfishness. Rather than thinking in terms of human 'nature', we should be thinking of human *potentialities* – and of how we can, collectively, build a society that is based on those positive and caring aspects associated with being human. Because it *is* possible: *if* everyone who's aware of one or more of the multiple crises we're facing, and 'who truly gets that we're living in a pivotal moment, found their place in the movement, amazing things could happen.'[17]

Probably the best-known song from the musical *Man from La Mancha* – a song of inspiration and unwavering determination – is the song 'The Impossible Dream', and it begins with the line: 'To dream the impossible dream.' While one of the more famous lines from Shakespeare's *Hamlet* is in Act II, Scene ii: '…there is nothing either good or bad, but thinking makes it so.' Given that Cervantes's *Don Quixote* was much admired by the revolutionary Che Guevara, perhaps it's high-time today's activists and revolutionaries combine those two lines, to conclude that 'tis only thinking it's an impossible dream makes it so'? That way, we can make sure we 'fight for the right, without question or pause' – because, if we do, we stand a *much* better chance of ensuring that 'the world will be better for this.' Just as Denmark was 'a prison' for Hamlet, those who feel a better, transformed, post-capitalist world is 'an impossible dream' are in effect isolated in capitalism's 'mind prison.' As said in the Introduction, to dream of the 'real possible' is not unfounded uto-pianism or 'hopium' – but realistic revolutionary dreaming. So: let's make it so!![18]

Yet, when confronted by the multiple crises we and the planet face, it's not surprising that many people – including some who're understandably 'burnt out' from taking incredible actions for the Earth to live, with organizations like Extinction Rebellion and Just Stop Oil – feel it is nonetheless an impossible dream. However – always bearing in mind our health, and caring and/or financial constraints – for things to improve, we do need to keep trying as much as we can: because if we don't, by far the most likely out-come is that capitalism's 'business-as-usual' will continue pushing the planet ever-forwards along its path of destruction of the nat-ural world. We can't afford to 'choose to look the other way… be-cause [we] seem more frightened of the changes that can prevent catastrophic climate change than the catastrophic climate change.' And, as Greta Thunberg said in her book *No One is Too Small to Make a Difference*: 'Every single person counts.' A similar idea is expressed by an ancient Chinese proverb, referenced in the 1998 animation film, *Mulan*, which has the 'Emperor' point out that it's always the last grain of rice that tips the scales, and that a single person may be the difference between victory and defeat. Anoth-er saying which makes the same point about the importance of individuals in making movements successful is this: 'every storm begins with a single raindrop.' So, if you're not already such a 'raindrop', you're invited to become one now! And so become part of a movement determined to bring about a better world founded on sustainability, justice and liberation for all.[19]

As historian Howard Zinn observed: 'history does not start anew with each decade. The roots of one era branch and flower in subsequent eras. Human beings, writings, invisible transmitters, carry messages across the generations.' Zinn was someone who, in his own words, was 'guilty' of a 'failure to quit' – a failure to quit the struggle for a better world, despite frequent setbacks. His attitude echoes that of Antonio Gramsci, whose views on pessimism and optimism in 'dark' times were referenced at the beginning of this book, and which have been mentioned again in this Conclusion. Zinn's more modern optimistic take on Gramsci's advice was simply this: 'I try to be pessimistic, to keep up with some of my friends. But I think back over the decades, and look around. And then it seems to me that the future is not certain, *but it is possible.*' And it is having a concrete and *realistic* vision of a 'Real Possible' ecosocialist future that, in the words of Ernest Mandel, 'has today become a prerequisite for practical-revolutionary activity.'[20]

'The longed-for tidal wave'

As stated at the beginning, this book is not aimed at those already on the ecosocialist 'page' of world history – though perhaps one or two sections may well stimulate further discussions amongst ecosocialist groups. Instead, this book was conceived as a project to be addressed to two main possible audiences: those involved in whatever way in the climate and environment movements who still do not yet see capitalism as the main cause of the multiple crises we face; and those in the labour movement who see capitalism as the problem, but who still do not see *eco*-socialism as the solution. In addition, it has been written *for* all activists who share at least some of the dreams and hopes expressed in the preceding pages – and particularly to encourage them to continue the struggle; and to encourage others to have the courage to take action to struggle for a better world. It has also been written *against* despair, depression and defeatism – because: 'Despair is paralysis. It robs us of agency. It blinds us to our own power … environmental despair is a poison.'[21]

Hopefully (pun intended!), by now at least some members of those two groups who've got this far are now convinced that, if only in broad outlines, the 'great transformation' we need to bring about on planet Earth *is* an ecosocialist one. If not, then maybe such readers can at least see the necessity to continue the struggle to achieve a more ecologically-sustainable and socially-just world – because taking part in struggles can bring about changes in the individuals involved in those struggles, as well as bringing about the changes needed to create that better world.

Sometimes, of course, those struggles will receive set – backs, which will test optimism and may lead to pessimism – but, as Bloch noted, 'a tested optimism, when the scales fall from the eyes, does not deny the goal-belief in general.' As has been argued earlier in this book, such *realistic* hope in the possibility of bringing about fundamental change – despite optimism being 'tested' at times by short-term failures – is crucial in all revolutionary struggles, because: 'Hope calls for action; action is impossible without hope ... To hope is to give yourself to the future, and that commitment to the future makes the present inhabitable.' Furthermore, as Solnit made clear: 'hope should shove you out the door, because it will take everything you have to steer the future away from...the annihilation of the earth's treasures.' Because, while building a future world that is better and fairer is definitely possible, it's 'not promised, not guaranteed.' Whilst it's impossible to predict that future, this book argues that what *is* predictable is that, in the absence of what Löwy has called a 'radical change in the civilizational paradigm' – or a fundamental ecosocialist transformation – 'the logic of capitalism will lead to dramatic ecological disasters, threatening the health and lives of millions of human beings and perhaps even the survival of our species.' So, for those in whom 'pessimism of the intellect' nonetheless still seems, at the moment, to be conquering the 'optimism of the will' urged upon us by Gramsci, perhaps these words, from the late and great Seamus Heaney, will help tip the balance the *right* way. Because sometimes, just when things seem hopelessly stuck, a turning point *is* reached:

> *History says, Don't hope*
> *On this side of the grave.*
> *But then, once in a lifetime*
> *The longed-for tidal wave*
> *Of justice can rise up,*
> *And hope and history rhyme.*[22]

ENDNOTES

Introduction: Optimism of the will

1. The Great Law of the Haudenosaunee of the Iroquois Confederacy *web.pdx. edu/~caskeym/iroquois_web/html/greatlaw.html* The second part of the 'Dedication' also comes from this site.

2. Karl Marx, *Capital Volume 3*, Harmondsworth, Penguin Books, 1981, p.911

3. Ruby Turner, 'Got To Be Done', from *Love Was Here*, RTR Productions, 2019: *www.youtube.com/watch?v=H3-5w1te8Qk*

4. *www.marxists.org/archive/gramsci/1924/03/pessimism.htm* Gramsci had helped launch this journal in 1919; but it seems the phrase was used originally by the French writer Romain Rolland.

5. Antonio Gramsci, *Selections from the Prison Notebooks*, London, Lawrence & Wishart, 1971, p.276

6. *www.centreforoptimism.com/Pessimism-of-the-Intellect-Optimism-of-the-Will*

7. Eric Hobsbawm, *Age of Extremes: The Short Twentieth Century 1914-1991*, London, Michael Joseph, 1994, p.585; Neil Faulkner, et al., *System Crash: An activist guide to making revolution*, London, Resistance Books, 2021, p.120

8. Allan Todd, *Che Guevara: The Romantic Revolutionary*, Barnsley, Pen & Sword, 2024, p.222

9. Ernst Bloch, E., *The Principle of Hope, Vol. 1*, Cambridge (Mass.), The MIT Press, 1995, p. 238; Vladimir I. Lenin, 'What is to be done?', 1902, in *Lenin: Selected Works, Vol. 1*, Moscow, Progress Publishers, 1975, pp.225-26; Ernest Mandel, 'We Must Dream', 1978, in *Ernest Mandel: Hope and Marxism*, (ed. Alex de Jong), London, Resistance Books/IIRE, 2022, p.124

10. Kate Soper, *Post-Growth Living: For an Alternative Hedonism*, London, Verso, 2020, p.4

11. Karl Marx & Friedrich Engels, *The Communist Manifesto*, Harmondsworth, Penguin Books, 1967, p.94; Karl Marx, 'Preface to A Contribution to the Critique of Political Economy', 1859, in Karl Marx, *Early Writings*, Harmondsworth, Penguin Books, 1975, p.426

12. Michael Löwy, *Ecosocialism: A Radical Alternative to Capitalist Catastrophe*, Chicago, Haymarket Book, 2015, p.xi

13. Friedrich Engels, *Dialectics of Nature*, London, Wellred Books, 2012, p.182 This was an unfinished work of 1883, but wasn't published until 1925.

14. Alan Thornett, *Facing the Apocalypse: Arguments for Ecosocialism*, London, Resistance Books, 2019, pp.40-41

15. Alan Thornett, ibid., 2019, p.44

16. Michael Löwy & Olivier Besancenot, *Revolutionary Affinities: Towards a Marxist-Anarchist Solidarity*, Oakland (California), PM Press, 2023, pp.152-53, emphasis added; *theanarchistlibrary.org/library/ lewis-herber-murray-bookchin-ecology-and-revolutionary-thought*

17. Rachel Carson, *Silent Spring*, London, Penguin Books, 2000, p.215

18. 18 Alan Thornett, op. cit., 2019, p.46-48

19. 19 Alan Thornett, ibid., 2019, pp.51-52

20. 20 *monthlyreview.org/2022/05/01/notes-on-exterminism-for-the-twenty-first-century-ecology-and-peace-movements/*

21. 21 Rudolf Bahro, *Socialism and Survival*, London, Heretic Books, 1982, p.24 (emphasis added)

22. 22 Alan Thornett, op. cit., 2019, pp.59-60; Raymond Williams, 'Socialism and Ecology', 1982, in Raymond Williams, ed. Robin Gale, *Resources of Hope: Culture, Democracy, Socialism*, London, Verso, 1989, pp. 210-11, 213-14, 225

23. 23 *fourth.international/en/world-congresses/57*

24. 24 *internationalviewpoint.org/spip.php?article5452*

25. 25 Simon Hannah, *Reclaiming the Future: A Beginner's Guide to Planning the Economy*, London, Pluto Press, 2024, p.2

Chapter 1: The climate shit we're in

1. David Attenborough, 03/12/2018, COP24, Katowice, Poland *unfccc.int/documents/185211*

2. John Bellamy Foster, *The Ecological Revolution*, New York, Monthly Review Press, 2009, p.44

3. *www.theguardian.com/environment/2024/apr/20/sunak-has-set-britain-back-on-net-zero-says-uks-climate-adviser*

4. Neil Faulkner et al., *System Crash: An activist guide to making revolution*, London, Resistance Books, 2021, pp.5-6; Michael J. Albert, *Navigating the Polycrisis: Mapping the Futures of Capitalism and the Earth*, Cambridge (Mass.), MIT Press, 2024, p. 1

5. Ulrich Brand & Markus Wissen, *The Imperial Mode of Living: Everyday Life and the Ecological Crisis of Capitalism*, London, Verso, 2021, p.xiii

6. *www.youtube.com/watch?v=OoZBpPnXY2c*; *www.thecitizen.org.au/articles/forget-2050-experts-say-its-2030-or-bust-for-net-zero-emissions*

7. Kate Soper, *Post-Growth Living: For an Alternative Hedonism*, London, Verso, 2020, p.11; Mark Lynas, *Our Final Warning: Six Degrees of Climate Emergency*, London, 4th Estate, 2020, p.x

8. *academic.oup.com/bioscience/article/73/12/841/7319571*

9. *academic.oup.com/bioscience/article/73/12/841/7319571*

10. *report-2023.global-tipping-points.org/summary-report/narrative-summary/*

11. *climate.copernicus.eu/warmest-december-concludes-warmest-year-record*; *global-tipping-points.org/summary-report/key-messages/*; Andreas Malm & Wim Carton, *Overshoot: How the World Surrendered to Climate Breakdown*, London, Verso, 2024, p.7

12. *news.un.org/en/story/2023/09/1140527 ; www.reuters.com/world/un-climate-chief-says-two-years-save-planet-2024-04-10/ ; phys.org/news/2024-01-extreme-east-antarctica-driven-atmospheric.html; www.theguardian.com/*

    environment/2024/apr/06/simply-mind-boggling-world-record-temperature-jump-in-antarctic-raises-fears-of-catastrophe

13. www.independent.co.uk/news/uk/home-news/farmers-warning-food-shortages-weather-b2526325.html; www.theguardian.com/world/2024/apr/18/dubai-floods-airport-malls-shopping-centre-uae-weather-record-rain

14. John Bellamy Foster et al., *The Ecological Rift: Capitalism's War on the Earth*, New York, Monthly Review Press, 2010, p.129; www.ucl.ac.uk/news/2020/jul/opinion-john-tyndall-forgotten-co-discoverer-climate-science

15. gml.noaa.gov/ccgg/trends/monthly.html https://www.co2.earth/daily-co2; John Bellamy Foster et al, op. cit., 2010, p.108

16. jksteinberger.medium.com/what-we-are-up-against-2290ba8c4b5c; www.nature.com/articles/s41893-022-00955-z

17. unric.org/en/guterres-the-ipcc-report-is-a-code-red-for-humanity/

18. anticapitalistresistance.org/the-earth-is-burning-the-earth-system-is-crumbling/

19. Naomi Klein, *This Changes Everything: Capitalism vs. the Climate*, London, Allen Lane, 2014, p.175; Mark Lynas, op. cit., 2020, p.ix; www.bbc.co.uk/news/science-environment-57487943

20. www.theguardian.com/environment/2023/nov/20/world-facing-hellish-3c-of-climate-heating-un-warns-before-cop28

21. climateactiontracker.org/global/temperatures/; climateactiontracker.org/publications/state-of-climate-action-2023/

22. www.noaa.gov/news/earth-had-its-hottest-august-in-175-year-record; www.bbc.co.uk/iplayer/episode/b00zj39j/frozen-planet-7-on-thin-ice

Chapter 2: The Sixth Mass Extinction

1. John Bellamy Foster et al., *The Ecological Rift: Capitalism's War on the Earth*, New York, Monthly Review Press, 2010, p.7

2. Foster et al., ibid., 2010, p.423; Edward O. Wilson, *Half-Earth: Our Planet's Fight for Life*, New York, Liveright Publishing Corporation, 2016, pp.8, 14, 59

3. ui.adsabs.harvard.edu/abs/2009Natur.461..472R/abstract

4. Kate Raworth, *Doughnut Economics: Seven Ways to Think Like a 21st.Century Economist*, London, Random House, 2017, p.76

5. Edward Wilson, op. cit., 2016, pp.1-2

6. Edward Wilson, ibid., 2016, p.3

7. Edward Wilson, ibid., 2016, pp.54-55

8. Edward Wilson, ibid., 2016, pp. 20, 56-57

9. Edward Wilson, ibid., 2016, p.60

10. Rachel Carson, *Silent Spring*, London, Penguin Books, 1999, pp.256-57

11. John Bellamy Foster, et al., op. cit., 2010, p.328

12. John Bellamy Foster, et al., ibid., 2010, p.329

13. John Bellamy Foster, et al., ibid., 2010, p.329

14. John Bellamy Foster, et al., ibid., 2010, p.330

15. Elizabeth Kolbert, *The Sixth Extinction: An Unnatural History*, London, Bloomsbury, 2015, pp.15-16; Chris Packham & A. Cohen, *Earth: Over 4 Billion Years in the Making*, London, William Collins, 2023, p.144

16. Elizabeth Kolbert, ibid., 2015, pp.17-18

17. Elizabeth Kolbert, ibid., 2015, pp.166-68

18. *news.bbc.co.uk/1/hi/sci/tech/3375447.stm; www.nationalgeographic. com/animals/article/110518-species-extinctions-habitats-science- animals; www.theguardian.com/environment/2023/nov/08/ species-at-risk-extinction-doubles-to-2-million-aoe*

19. Elizabeth Kolbert, op. cit., 2015, pp.171-2

20. Elizabeth Kolbert, ibid., 2015, p.189

21. *www.unep.org/news-and-stories/press-release/ natures-dangerous-decline-unprecedented-species-extinction-rates*

22. *www.unep.org/news-and-stories/press-release/ natures-dangerous-decline-unprecedented-species-extinction-rates*

23. John Bellamy Foster, et al., op. cit., 2010, p.271

24. John Bellamy Foster, et al., ibid., 2010, p.128

25. Elizabeth Kolbert, op. cit., 2015, p.108; *www.youtube.com/ watch?v=2Jq23mSDh9U*

26. Elizabeth Kolbert, ibid., 2015, p.113; *www.youtube.com/ watch?v=DZjsJdokC0s*

27. *climateandcapitalism.com/2024/03/13/why-the-anthropocenes-critics- are-wrong/; www.episodes.org/journal/view.html?doi=10.18814/ epiiugs/2023/023025*

28. Edward Wilson, op. cit., 2016, p.59

29. *climateandcapitalism.com/2024/03/13/why-the-anthropocenes-critics- are-wrong/; www.episodes.org/journal/view.html?doi=10.18814/ epiiugs/2023/023025*

30. Elizabeth Kolbert, op cit., 2015, p.114

31. Elizabeth Kolbert, ibid., 2015, p.123

32. Elizabeth Kolbert, ibid., 2015, pp.123-4

33. Edward Wilson, op. cit., 2016, p.4

34. John Bellamy Foster & Paul Burkett, *Marx and the Earth: An Anti-Critique*, Chicago, Haymarket Books, 2017, p.6

Chapter 3: The threat from pandemics

1. *jembendell.com/2020/03/23/ the-climate-for-corona-our-warming-world-is-more-vulnerable-to-pandemic/*

2. John B. Foster, *The Return of Nature: Socialism and Ecology*, New York, Monthly Review Press, 2020, p.7; Michael Löwy, *Ecosocialism: A Radical Alternative to Capitalist Catastrophe*, Chicago, Haymarket Books, 2015, p.vii

3.   Friedrich Engels, *Dialectics of Nature*, Wellred Books, London, 2012, pp.182-83; *www.youtube.com/watch?v=RMhbr2XQblk*

4.   *climateandcapitalism.com/2024/03/26/capitalisms-new-age-of-plagues-part-3/*; Michael Greger, *How to Survive a Pandemic*, London, Bluebird Books, 2020, p.xvi

5.   *www.who.int/data/stories/the-true-death-toll-of-covid-19-estimating-global-excess-mortality*; *ourworldindata.org/excess-mortality-covid*

6.   *www.nature.com/articles/nature.2014.14723*; Michael Greger, op. cit., 2020, pp.xi-xii, xv

7.   *news.mongabay.com/2020/04/ rapid-deforestation-of-brazilian-amazon-could-bring-next-pandemic-experts/*

8.   Giorgos Kallis, et al., *The Case for Degrowth*, Cambridge, Polity, 2020, pp.vii-viii

9.   Neil Faulkner, et al., *System Crash: an activist guide to making revolution*, London, Resistance Books, 2021, p.13

10. Neil Faulkner, et al., ibid., 2021, p.13; *metro.co.uk/2021/07/14/over-1500-nhs-and-care-workers-have-now-died-with-covid-14930073/*; *www.telegraph.co.uk/news/2020/03/28/exercise-cygnus-uncovered-pandemic-warnings-buried-government/*; *www.theguardian.com/world/2020/may/07/ what-was-exercise-cygnus-and-what-did-it-find*

11. Neil Faulkner, et al., ibid., 2021, p10

12. Michael Löwy, 'What is Ecosocialism?', 2005, in Jane Kelly & Sheila Malone, (eds), *Ecosocialism or barbarism*, London, Resistance Books, 2006, p.1; I. Angus, 'The Discovery and Rediscovery of Metabolic Rift', in Martin Empson (ed.), *System Change not Climate Change: A Revolutionary Response to Environmental Crisis,* London, Bookmarks Publications, 2019, p.51. (emphasis added)

13. Michael Löwy, op.cit., 2015, pp.84-85; *climateandcapitalism. com/2018/07/22/the-belem-ecosocialist-declaration-an-historic-document/*

14. Paul Burkett, *Marx and Nature: A Red and Green Perspective*, Chicago, Haymarket Books, 2014, p.xxi; Andreas Malm, *The Progress of this Storm: Nature and Society in a Warming World*, London, Verso, 2018, p.177

15. John B. Foster & Paul Burkett, *Marx and the Earth: An Anti-Critique*, Chicago, Haymarket Books, 2017, p.6. Neil Faulkner, et al., op. cit., 2021, p.11

16. Grace Blakeley, *Vulture Capitalism: Corporate Crimes, Backdoor Bailouts and the Death of Freedom*, London, Bloomsbury Publishing, 2024, p.85; *monthlyreview.org/2020/05/01/covid-19-and-circuits-of-capital/*

17. Allan Todd, *Che Guevara: The Romantic Revolutionary*, Barnsley, Pen & Sword, 2024, p.60; Neil Faulkner, et al., op. cit., 2021, p.16

18. Giorgos Kallis, et al., 2020, pp.vi-vii

19. Sheila Malone, 'Kyoto's answer to climate change', in Jane Kelly & Sheila Malone, (eds), op. cit., 2006, p.29

20. *anticapitalistresistance.org/capitalisms-new-age-of-plagues-part-5-the-pandemic-machines/ (This link also gives access to Parts 1-4 of this series)*; Rob Wallace, et al., 'COVID-19 and Circuits of Capital', in *Monthly Review 72, No. 1,* 1 May 2020, *monthlyreview.org/2020/05/01/*

covid-19-and-circuits-of-capital/

21. *socialistresistance.org/covid-19-the-ecological-dimension/19641* (emphasis added)

22. Fred Magdoff & Chris Williams, *Creating an Ecological Society: Toward a Revolutionary Transformation*, New York, Monthly Review Press, 2017, p.255

23. Alan Thornett, *Facing the Apocalypse: Arguments for Ecosocialism*, London, Resistance Books, 2019, p.173

24. *www.fao.org/4/a0701e/a0701e00.htm*

25. Alan Thornett, op. cit., 2019, p.186

26. Nancy Klein, *The Shock Doctrine: The Rise of Disaster Capitalism*, London, Penguin Books, 2008, pp.9, 13, 15; Neil Faulkner, et. al., op. cit., 2021, p.11

27. Neil Faulkner, et al., ibid., 2021, p.12; *news.sky.com/story/amp/coronavirus-life-in-barrow-in-furness-the-town-with-the-highest-infection-rate-in-the-country-11989473*

28. Giorgos Kallis, et al, op. cit., 2020, p.vii

29. Michael Greger, op. cit., 2020, p.x

30. *www.standard.co.uk/news/world/what-is-disease-x-vaccine-pandemic-wef-davos-b1099041.html; monthlyreview.org/2020/05/01/covid-19-and-circuits-of-capital/*

31. *www.bbc.com/future/article/20240425-how-dangerous-is-bird-flu-spread-to-wildlife-and-humans; www.cdc.gov/bird-flu/situation-summary/index.html; www.bbc.com/future/article/20240425-how-dangerous-is-bird-flu-spread-to-wildlife-and-humans; www.npr.org/sections/shots-health-news/2024/07/24/nx-s1-5049893/u-s-bird-flu-outbreak-scientists-see-growing-risks*

32. *www.theguardian.com/environment/2024/apr/25/mosquito-borne-diseases-spreading-in-europe-due-to-climate-crisis-says-expert*

33. John B. Foster, op. cit., 2020, p.7; Neil Faulkner, et. al., op. cit., 2021, p.14

34. Giorgos Kallis, et al, op. cit., 2020, pp xix-xi

35. Michael Löwy, op.cit., 2015, pp.1-2

36. Ian Angus, *Facing the Anthropocene: Fossil Capitalism and the Crisis of the Earth System*, New York, Monthly Review Press, 2016, p.191

Chapter 4: The twilight of democracy

1. Clara Mattei, *The Capital Order: How Economists Invented Austerity and Paved the Way to Fascism*, London, University of Chicago Press, 2022, p.164 (emphasis in original)

2. Neil Faulkner, et al., *System Crash: an activist guide to making revolution*, London, Resistance Books, 2021, p.62

3. *jksteinberger.medium.com/what-we-are-up-against-2290ba8c4b5c*

4. George Monbiot & Peter Hutchison, *The Invisible Doctrine: The Secret History of Neoliberalism*, Allen Lane, 2024, p.16

5. George Monbiot & Peter Hutchison, ibid., 2024, p.19

6.   *jksteinberger.medium.com/what-we-are-up-against-2290ba8c4b5c*

7.   *jksteinberger.medium.com/what-we-are-up-against-2290ba8c4b5c*

8.   Tariq Ali, *The Extreme Centre: A Warning*, London, Verso, 2015, pp.1, 10, 15; Paul Mason, *How to Stop Fascism: History, Ideology, Resistance*, London, Allen Lane, 2021, p.6

9.   *x.com/cpsuk/status/862245330745790464; www.heraldscotland.com/ news/15276477.no-charges-will-brought-tory-spending-2015-election-rules-crown-prosecution-service/*

10. *www.youtube.com/watch?v=6I_ZhGHxnHQ; www.doubledown.news/watch/2020/30/june/ george-monbiot-on-tory-corruption-robert-jenrick-richard-desmond*

11. *www.independent.co.uk/news/uk/politics/supreme-court-decision-ruling-boris-johnson-suspension-prorogue-brexit-latest-today-a9117931.html*

12. *www.libertyhumanrights.org.uk/issue/libertys-view-on-voter-id-plans-opportunistic-divisive-and-undemocratic/; uk.news.yahoo.com/voter-id-expansion-urged-750k-094221528.html*

13. *anticapitalistresistance.org/cass-means-social-murder/; anticapitalistresistance.org/strange-playgrounds-the-cass-report-and-the-uk-trans-panic-with-rowan-fortune/*

14. Grace Blakeley, *Vulture Capitalism: Corporate Crimes, Backdoor Bailouts and the Death of Freedom*, London, Bloomsbury Publishing, 2024, p.5

15. Clara Mattei, op. cit., 2022, pp.3, 14

16. *www.independent.co.uk/news/health/tory-austerity-deaths-study-report-people-die-social-care-government-policy-a8057306. html www.theguardian.com/business/2022/oct/05/ over-330000-excess-deaths-in-great-britain-linked-to-austerity-finds-study*

17. *www.trusselltrust.org/news-and-blog/latest-stats/end-year-stats/*

18. *www.theguardian.com/environment/2022/jun/29/five-things-we-have-learned-about-the-uks-path-to-net-zero; www.ft.com/ content/24117a03-37c2-424a-97ed-6a5292f9e92e*

19. *www.forbes.com/sites/chasewithorn/2024/04/02/ forbes-38th-annual-worlds-billionaires-list-facts-and-figures-2024/*

20. *www.theguardian.com/inequality/2024/apr/25/billionaires-should-pay-minimum-two-per-cent-wealth-tax-say-g20-ministers; www.greenpeace.org. uk/news/economists-campaigners-millionaires-calls-for-wealth-tax/; www. greenpeace.org.uk/news/national-renewal-tax-on-the-super-rich-report/*

21. Tariq Ali, op. cit., 2015, London, Verso, 2015, p.12

22. Neil Faulkner, et al., op. cit., 2021, p.62; *www.youtube.com/ watch?v=RMr2R_67wL0*

23. *anticapitalistresistance.org/ tory-government-continues-its-attack-on-civil-rights/*

24. *www.theguardian.com/environment/2021/aug/31/ police-wield-batons-during-xrs-london-bridge-bus-blockade?fr=operanews*

25. *www.theguardian.com/commentisfree/2021/oct/01/ boris-johnson-rigging-the-system-power-courts-protest-elections*

26. Naomi Klein, *This Changes Everything: Capitalism vs. the Climate*, London, Allen Lane, 2014, pp.8-9

27. Kate Soper, *Post-Growth Living: For an Alternative Hedonism*, London, Verso, 2020, p.40; *www.theguardian.com/commentisfree/2018/oct/18/governments-no-longer-trusted-climate-change-citizens-revolt; www.theguardian.com/news/2022/sep/04/super-rich-prepper-bunkers-apocalypse-survival-richest-rushkoff?ck_subscriber_id=2034845217*

28. *theintercept.com/2020/06/05/pentagon-war-game-gen-z/*

29. *www.theguardian.com/uk-news/2024/mar/07/socialism-anti-fascism-anti-abortion-prevent-list-terrorism-warning-signs?CMP=share_ btn_url; www.theguardian.com/uk-news/article/2024/may/21/ new-police-powers-for-protests-unlawful-high-court-rules*

30. *jksteinberger.medium.com/what-we-are-up-against-2290ba8c4b5c*

Chapter 5: Creeping fascism

1. Neil Faulkner, et al., *System Crash: an activist guide to making revolution*, London, Resistance Books, 2021, p.60

2. *thenextrecession.wordpress.com/2024/11/09/us-election-2024-inflation-immigration-and-identity/; www.bbc.co.uk/news/articles/c977njnvq2do*

3. *jksteinberger.medium.com/what-we-are-up-against-2290ba8c4b5c*

4. Neil Faulkner, et al., *Creeping Fascism: What it is and how to fight it*, London, Public Reading Rooms, 2019, p.22

5. Leon Trotsky, 'Whither France?', 1934, in *Leon Trotsky on France*, New York, Monad Press, 1979, p.39; Leon Trotsky, 'What Next? Vital Questions for the German Proletariat', 1932, in Leon Trotsky, *The Struggle Against Fascism in Germany*, New York, Pathfinder Press, 1971, pp. 155,192

6. Leon Trotsky, 'For a Workers' United Front Against Fascism', 8 December 1931, in Leon Trotsky, op. cit., 1971, p.141

7. Paul Mason, *How to Stop Fascism: History, Ideology, Resistance*, London, Allen Lane, 2021, p.194

8. Paul Mason, ibid., 2021, p.75

9. *www.bbc.co.uk/news/uk-66890135; medicalxpress.com/news/2024-05-reveals-people-died-hot-cold.html; www.nature.com/articles/ s41467-024-48207-2*

10. *jksteinberger.medium.com/what-we-are-up-against-2290ba8c4b5c*

11. Neil Faulkner, et al., op. cit., 2021, p.63

12. Paul Stocker, *English Uprising: Brexit and the Mainstreaming of the Far Right*, London, Melville House, 2017, p.208

13. *amp.theguardian.com/uk-news/2018/jul/22/ us-rightwing-groups-bankroll-campaign-to-free-tommy-robinson*

14. Paul Mason, op. cit., 2021, pp.194, 215, 225

15. *jksteinberger.medium.com/what-we-are-up-against-2290ba8c4b5c; link. springer.com/book/10.1007/978-981-15-3936-7*

16. Mike Wendling, *Alt-Right: From 4chan to the White House*, London, Pluto Press, 2018, p.3; Neil Faulkner, et. al., op. cit., 2021, p.68; Paul Mason, op. cit., 2021, p.xv

17. Neil Faulkner, et al., op. cit., 2019, p.312; Neil Faulkner, et. al., op. cit., 2021, pp.70-1; *anticapitalistresistance.org/creeping-fascism-revisited/*

18. Enzo Traverso, *The New Faces of Fascism: Populism and the Far Right*, London, Verso, 2019, pp.27-8 (emphasis in the original)

19. Neil Faulkner, *Mind Fuck: The Mass Psychology of Creeping Fascism*, London, Resistance Books, 2022, p.116

20. Paul Stocker, op. cit., 2017, p.111; *Searchlight, www.searchlightmagazine. com/2023/07/gateway-to-the-far-right/#*

21. Paul Stocker, op. cit., 2017, pp.176-7

22. Darren McGarvey, *Poverty Safari: Understanding the anger of Britain's Underclass*, London, Picador, 2018, p.xx

23. McGarvey, ibid., 2018, pp.36-7, 48

24. *www.youtube.com/shorts/bjs144juNK4*

25. Enzo Traverso, op. cit., 2019, p.5; Neil Faulkner, et. al., op. cit., 2019, p.8

26. Neil Faulkner, et. al., ibid., 2019, p. 120

27. *news.sky.com/story/robert-jenrick-profile-who-is-the-tory-leadership-candidate-that-went-from-centrist-remainer-to-trump-backing-voice-of-the-right-13238622; www.nytimes.com/2024/11/02/world/europe/kemi-badenoch-uk-tory-leadership-election-2024.html*

28. Neil Faulkner, et. al., ibid., 2019, pp. 32, 120, 298

29. *www.desmog.com/2024/06/04/nigel-farage-reform-uk-party-2-3-million-fossil-fuel-interests-climate-deniers-polluters-since-2019-election/; www. desmog.com/2024/09/20/mapped-reform-uk-nigel-farage-anti-green-network/*

30. Paul Mason, op. cit., 2021, pp.200, 214-15

31. Paul Mason, ibid.., 2021, pp.xxii, 74; Neil Faulkner, et al., op. cit., 2021, p.70

32. Neil Faulkner, et. al., op. cit., 2019, pp.314, 322

33. *www.versobooks.com/en-gb/blogs/news/1683-ten-theses-on-the-far-right-in-europe-by-michael-lowy*; Neil Faulkner, et. al., op. cit., 2019, p.314. (emphasis in the original)

34. *www.reuters.com/world/europe/how-far-right-gained-traction-with-europes-youth-2024-06-13/; anticapitalistresistance.org/european-elections-far-right-surge-but-centre-holds-on/; link.springer.com/article/10.1007/s43638-023-00078-y*

35. *anticapitalistresistance.org/sacre-bleu-rouge-popular-front-pushes-back-le-pen/; anticapitalistresistance.org/voters-didnt-vote-for-that/*

36. *inews.co.uk/news/politics/reform-poll-surge-continues-warning-tories-labour-3309571; www.youtube.com/watch?v=bjs144juNK4; anticapitalistresistance.org/solidarity-and-unity-now/*; Paul Gale, 'A Shifting

Kaleidoscope: The State of Britain's Far Right in 2024', in Searchlight, No. 495, Ilford, Autumn 2024, p.9

Chapter 6: Blood and fire: Birth of a monster

1. Karl Marx, *Capital Volume 1*, Harmondsworth, Penguin Books, 1976, p.875

2. Karl Marx, ibid., 1976, p.928

3. Ian Angus, *The War Against the Commons: Dispossession and Resistance in the Making of Capitalism*, New York, Monthly Review Press, 2023, p.10

4. Karl Marx, *Capital Volume 3*, Harmondsworth, Penguin Books, 1981, p.571; Karl Marx, op. cit., 1976, p.926

5. John Clare, 'The Mores', (1820), in Summerfield, G. (ed.), *Selected Poems*, London, Penguin Books, 1990, pp. 170-71

6. Karl Marx, op. cit., 1976, pp.874, 876; Ian Angus, op. cit., 2023, p.11

7. Ian Angus, ibid., 2023, pp.26-27

8. Ian Angus, ibid., 2023, p.31

9. Matthew Champion & Nicholas Sotherton, *Kett's Rebellion 1549*, Kerdiston (Norfolk), Timescape Publishing, 1999, p.97

10. Anthony Fletcher & Diarmaid MacCulloch, *Tudor Rebellions*, Harlow, Pearson Education, 2004, p.70; Ian Angus, op. cit., 2023, p.19; Karl Marx, op. cit., 1976, p.886

11. A. L. Beier, *The problem of the poor in Tudor and Stuart England*, London, Routledge, 1983, p.9

12. W. Shenk, London, Longmans, Green & Co., 1948, p.78; Bernstein, *Cromwell and communism, socialism & democracy in the great English revolution*, London, George Allen & Unwin, 1930, pp.105-06

13. L. H. Berens, 1906, *The Digger Movement in the Days of the Commonwealth*, London, Holand Press & Merlin Press, p.229

14. Crawford B. Macpherson, *The Political Theory of Possessive Individualism: Hobbes to Locke*, Oxford, Clarendon Press, 1962, p.3

15. Ian Angus, op. cit., 2023, p.47

16. Karl Marx, op. cit., 1976, p.876

17. Ian Angus, op. cit., 2023, p.59

18. Karl Marx, op. cit., 1976, pp.885-86

19. Karl Marx & Friedrich Engels, *The Communist Manifesto*, Harmondsworth, Penguin Books, 1967, p.98

20. Ian Angus, op. cit., 2023, pp.177-81

21. Ian Angus, ibid., 2023, pp.179-81

22. John B. Foster & Brett Clark, *The Robbery of Nature: Capitalism and the Ecological Rift*, New York, Monthly Review Press, 2020, p.13; Karl Marx, op. cit., 1976, pp.637-38

23. Karl Marx, op. cit., 1981, p.949; John B. Foster & Brett Clark, ibid., 2020, pp.120-7

24. John B. Foster & Brett Clark, ibid., 2020, pp.117, 119; Ian Angus, op. cit., 2023, pp.162, 166

25. Ian Angus, ibid., 2023, pp.162-63, 167; John B. Foster & Brett Clark, ibid., 2020, p.118; *www.sciencedirect.com/science/article/pii/S0305750X22002169* (emphasis added)

26. John B. Foster & Brett Clark, ibid., 2020, pp.21, 120, 126

27. John B. Foster & Brett Clark, ibid., 2020, p.26

28. John B. Foster & Brett Clark, ibid., 2020, pp. 23, 127-8; Karl Marx, op. cit., 1976, p.638

29. David Whyte, *Ecocide: Kill the Corporation Before it Kills Us*, Manchester, Manchester University Press, 2020, pp.79-80

30. David Whyte, ibid., 2020, p.83

31. Mikaela Loach, *It's Not That Radical: Climate Action to Transform Our World*, London, Dorling Kindersley Ltd., 2024, p. 55

32. *www.brh.org.uk/site/book-reviews/the-blood-never-dried-a-peoples-history-of-the-british-empire/* ; Mikaela Loach, ibid., 2024, p.57

33. *www.sciencedirect.com/science/article/pii/S0305750X22002169*

34. Rosa Luxemburg, *The Accumulation of Capital*, London, Routledge & Kegan Paul 1971, p.369

35. *www.sciencedirect.com/science/article/pii/S0305750X22002169*

36. Ian Angus, op. cit., 2023, p.164

37. David Whyte, op. cit., 2020, p.81

38. Mikaela Loach, ibid., 2024, p.55; Rosa Luxemburg, op. cit., 1971, pp.369-70

39. *jksteinberger.medium.com/what-we-are-up-against-2290ba8c4b5c; https://www.jasonhickel.org/the-divide*

40. *https://www.jasonhickel.org/the-divide; jksteinberger.medium.com/what-we-are-up-against-2290ba8c4b5c*

Chapter 7 – Out of control

1. Kate Raworth, *Doughnut Economics: Seven Ways to Think Like a 21st-Century Economist*, London, Random House, 2018, p.212

2. Karl Marx & Friedrich Engels, *The Communist Manifesto*, Harmondsworth, Penguin Books, 1967, p.85; *monthlyreview.org/2004/10/01/capitalism-and-the-environment/*

3. John B. Foster, *The Ecological Revolution: Making Peace with the Planet*, New York, Monthly Review Press, 2009, pp.40-41; *monthlyreview.org/product/socialism_or_barbarism/*

4. *monthlyreview.org/2004/10/01/capitalism-and-the-environment/*

5. John B. Foster, op. cit., 2009, pp.57-58

6. John B. Foster, ibid., 2009, p.58; John B. Foster, et al., *The Ecological Rift: Capitalism's War on the Earth*, New York, Monthly Review Press, 2010, p.108; *350.org/about/*

7. John B. Foster, ibid., 2009, p.58

8. *journals.openedition.org/sapiens/240*

9. John B. Foster, op. cit., 2009, pp.59-60

10. John B. Foster, ibid., 2009, p.57 (emphasis in the original)

11. John B. Foster, ibid., 2009, p.59; *www.eea.europa.eu/en/analysis/indicators/atmospheric-greenhouse-gas-concentrations*

12. *www.ipcc.ch/report/ar6/syr/summary-for-policymakers/*

13. *wmo.int/news/media-centre/greenhouse-gas-concentrations-hit-record-high-again*

14. John B. Foster et al., op. cit., 2010, p.109; *press.un.org/en/2022/sgsm21228.doc.htm*

15. *www.science.org/doi/10.1126/sciadv.1500589*

16. *www.vox.com/climate/2024/2/28/24085691/atlantic-ocean-warming-climate-change-hurricanes-coral-reefs-bleaching*; *www.nature.com/articles/s41558-023-01919-7*

17. *news.un.org/en/story/2024/06/1150661*; Andreas Malm & Wim Carton, *Overshoot: How the World Surrendered to Climate Breakdown*, London, Verso, 2024, p.21 (emphasis added)

18. *www.eea.europa.eu/en/analysis/indicators/atmospheric-greenhouse-gas-concentrations* ; *news.un.org/en/story/2024/06/1150661*

19. *www.theguardian.com/environment/ng-interactive/2024/may/08/hopeless-and-broken-why-the-worlds-top-climate-scientists-are-in-despair*; *www.theguardian.com/environment/article/2024/may/08/world-scientists-climate-failure-survey-global-temperature*

20. *www.carbonbrief.org/state-of-the-climate-2024-off-to-a-record-warm-start/*

21. *www.bbc.co.uk/news/science-environment-68921215*

22. *www.theguardian.com/environment/article/2024/may/11/brutal-heatwaves-submerged-cities-what-3c-world-would-look-like*

23. *climate.copernicus.eu/copernicus-june-2024-marks-12th-month-global-temperature-reaching-15degc-above-pre-industrial*; *www.theguardian.com/world/article/2024/may/22/never-ending-uk-rain-10-times-more-likely-climate-crisis-study*; *www.bbc.co.uk/news/articles/cn4zy0mp8ldo*

24. *www.theguardian.com/world/article/2024/may/29/delhi-temperature-hits-499c-as-indias-capital-records-hottest-day*; *www.unocha.org/publications/report/bangladesh/bangladesh-eastern-flash-floods-2024-situation-report-no-02-30-august-2024*

25. *www.weforum.org/videos/how-16-tipping-points-could-push-our-entire-planet-into-crisis/* (emphasis added); *academic.oup.com/oocc/article/3/1/kgad008/7335889*; *www.stockholmresilience.org/research/research-videos/2024-08-19-the-tipping-points-of-climate-change---and-where-we-stand.html*

26. *www.theguardian.com/world/2024/oct/03/wildfires-are-burning-through-humanitys-carbon-budget-study-shows?CMP=share_btn_url*

27. *www.bbc.co.uk/news/articles/c89vw50kj3yo* ; *www.bbc.co.uk/news/articles/cz0m7exe53do*; *edition.cnn.com/2024/10/31/asia/taiwan-typhoon-kong-rey-*

landfall-intl-hnk/index.html; www.reuters.com/world/europe/heavy-rains-cause-flash-floods-spains-south-east-2024-10-29/; www.theguardian.com/world/2024/nov/02/spain-apocalyptic-floods-climate-crisis-worse-big-oil-cop29; www.bbc.co.uk/news/live/cgk1m7g73ydt; www.reuters.com/business/environment/climate-set-warm-by-31-c-without-greater-action-un-report-warns-2024-10-24/

28. *pulse.climate.copernicus.eu/; www.theguardian.com/environment/2024/nov/07/this-year-virtually-certain-to-be-hottest-on-record-finds-eu-space-programme; wmo.int/news/media-centre/2024-track-be-hottest-year-record-warming-temporarily-hits-15degc; www.bbc.co.uk/news/articles/crmzvdn9e18o; www.bbc.co.uk/news/articles/cwyx0xw5vyyo*

Chapter 8 – Deadly paradoxes and killer corporations

1.   Grace Blakeley, *Vulture Capitalism: Corporate Crimes, Backdoor Bailouts and the Death of Freedom*, London, Bloomsbury Publishing, 2024, p.161

2.   John B. Foster et al., *The Ecological Rift: Capitalism's War on the Earth*, 2010, pp.139, 141; John B. Foster, *The Ecological Revolution: Making Peace with the Planet*, 2009, pp.125-26

3.   John B. Foster et al., ibid., 2010, pp.139-41 (emphasis in original)

4.   John B. Foster et al., ibid., 2010, pp.140-42, 179-80,185-86

5.   *www.vox.com/climate/363076/climate-change-solution-shell-exxon-mobil-carbon-capture*; Daniel Tanuro, *Green Capitalism: why it cannot work*, London, Merlin Press/IIRE, 2013, pp.104-05

6.   John B. Foster & Brett Clark, *The Robbery of Nature: Capitalism and the Ecological Rift*, New York, Monthly Review Press, 2020, p.154; *Ecosocialist Revolution: A Manifesto*, London, AntiCapitalist Resistance, 2024, pp.15-16

7.   John B. Foster & Brett Clark, ibid., 2020, pp.154, 157

8.   John B. Foster & Brett Clark, ibid., 2020, pp.158-59; John B. Foster et al., ibid., 2010, pp.433-54

9.   Karl Marx, *Capital Volume 1*, Harmondsworth, Penguin Books,1976, pp.133-34, 381; John B. Foster & B. Clark, ibid., 2020, p.163 (emphasis in the original)

10. John B. Foster & Brett Clark, ibid., 2020, pp.167-69

11. John B. Foster & Brett Clark, ibid., 2020, pp.169-72

12. *www.independent.co.uk/climate-change/news/bp-shell-oil-global-warming-5-degree-paris-climate-agreement-fossil-fuels-temperature-rise-a8022511.html; www.independent.co.uk/climate-change/news/arctic-circle-sea-ice-free-global-warming-limit-two-degrees-celcius-climate-change-paris-agreement-targets-a7616311.html*

13. Mark Lynas, M., *Our Final Warning: Six Degrees of Climate Emergency*, London, 4th. Estate, 2020, p.215; John B. Foster et al., op. cit., 2010, p.109

14. David Whyte, *Ecocide: Kill the Corporation Before it Kills Us*, Manchester, Manchester University Press, 2020, p.3

15. Grace Blakeley, op. cit., 2024, pp. 7, 98, 120-21

16. *monthlyreview.org/2004/10/01/capitalism-and-the-environment/*; David Whyte, op. cit., 2020, p.2

17. David Whyte, ibid., 2020, pp.80, 85; *jksteinberger.medium.com/what-we-are-up-against-2290ba8c4b5c*

18. David Whyte, ibid., 2020, pp.25-27, 69, 87

19. David Whyte, ibid., 2020, pp.28, 39, 69; Mikaela Loach, *It's Not That Radical: Climate Action to Transform Our World,* London, Dorling Kindersley Ltd., 2024, pp.55-56

20. David Whyte, ibid., 2020, pp.3, 21, 80; Grace Blakeley, op. cit., 2024, p.107; *www.youtube.com/watch?v=RvAOuhyunhY*

21. Karl Marx & Friedrich Engels, *The Communist Manifesto,* Harmondsworth, Penguin Books, 1967, p.82; Daniel Guérin, *Fascism and Big Business,* New York, Monad Press, 1973, p.21; Grace Blakeley, ibid., 2024, pp.3, 5

22. George Monbiot & Peter Hutchison, *The Invisible Doctrine: The Secret History of Neoliberalism,* Allen Lane, 2024, p.51

23. George Monbiot & Peter Hutchison, ibid., 2024, p.51

24. George Monbiot & Peter Hutchison, ibid., 2024, p.53; *freedomhouse.org/article/new-report-global-decline-democracy-has-accelerated; freedomhouse.org/report/freedom-world/2024/mounting-damage-flawed-elections-and-armed-conflict*

25. *jacobin.com/2023/06/democracy-retreat-capitalism-authoritarianism-crisis?mc_cid=94e2e5a06c* ; Grace Blakely, op. cit., 2024, p.160

26. *www.pressreader.com/australia/the-guardian-austral ia/20240322/2825186663491766* ; Grace Blakely, ibid., 2024, pp.159-60. (emphasis added)

27. Grace Blakely, ibid., 2024, p. 160. (emphasis in the original); Kate Raworth, *Doughnut Economics: Seven Ways to Think Like a 21st-Century Economist,* London, Random House Business Books, 2018, p.41

28. Kate Raworth, ibid., 2018, pp.5-7, 287 (emphasis added); *issuu.com/dartmouth_college_library/docs/the_limits_to_growth?e=1347206/1573259; www.youtube.com/watch?v=hhSpzQhvFS8*

29. *monthlyreview.org/2004/10/01/capitalism-and-the-environment/*; John B. Foster et al., ibid., 2010, pp.157-9; Jason Hickel, *Less is More: How Degrowth Will Save the World,* London, Penguin Books, 2020, pp.81, 83

Chapter 9: In greater harmony with nature

1. Ian Angus, *The War Against the Commons: Dispossession and Resistance in the Making of Capitalism,* New York, Monthly Review Press, 2023, p.9

2. *climateandcapitalism.com/2012/11/14/who-defends-earths-legacy/*

3. *www.sciencedirect.com/science/article/abs/pii/S0305440306002019;* www.discovermagazine.com/planet-earth/easters-end

4. *www.sciencedirect.com/science/article/abs/pii/S0305440320300182?via%3Dihub; www.sapiens.org/archaeology/easter-island-collapse/*

5.   *www.smithsonianmag.com/science-nature/why-did-the-mayan-civilization-collapse-a-new-study-points-to-deforestation-and-climate-change-30863026/*

6.   *www.livescience.com/why-maya-civilization-collapsed.html*

7.   Karl Marx, *Capital Volume 1*, Harmondsworth, Penguin Books, 1976, p.273

8.   Kate Raworth, *Doughnut Economics: Seven Ways to Think Like a 21st-Century Economist*, London, Random House Business Books, 2018, pp.82-83; J. M. Neeson, *Commoners: Common Right, Enclosure and Social Change in England, 1700-1820*, Cambridge, Cambridge University Press, 1993, pp. 112-13

9.   Ian Angus, op. cit, 2023, p.23; Derek Wall, *Elinor Ostrom's Rules for Radicals: Cooperative Alternatives Beyond Markets and States*, London, Pluto Press, 2017, pp.32-33, 36

10. Ian Angus, ibid., 2023, p.24

11. *www.bbc.co.uk/iplayer/episode/m0022710/brian-may-the-badgers-the-farmers-and-me*; Ian Angus, op. cit., 2023, p.24

12. Karl Marx, *Capital Volume 3*, Harmondsworth, Penguin Books, 1981, p.959

13. *ia800201.us.archive.org/32/items/green-entrepreneurship-2021/Hardin%20%281968%29%20Science%20-%20The%20tragedy%20of%20the%20Commons.pdf*; Kate Raworth, op. cit., 2018, p.83

14. *ia800201.us.archive.org/32/items/green-entrepreneurship-2021/Hardin%20%281968%29%20Science%20-%20The%20tragedy%20of%20the%20Commons.pdf*; Kate Raworth, ibid., 2018, p.83; Ian Angus, op. cit., 2023, p.21

15. Ian Angus, ibid., 2023, pp.22-23; *socialsciencelibrary.org/philosophy/the-environment/environmental-politics/no-tragedy-of-the-commons/*

16. *www.sciencedirect.com/science/article/pii/S0305750X22002169*

17. Kate Raworth, op. cit., 2018, pp.83, 181; Derek Wall, op. cit., 2017, p.2

18. Derek Wall, ibid., 2017, p.35; *www.project-syndicate.org/commentary/green-from-the-grassroots-2012-06*

19. Derek Wall, ibid., 2017, pp.36, 40, 43

20. Rosa Luxemburg, *The Accumulation of Capital*, London, Routledge & Kegan Paul, 1971, p.368

21. Derek Wall, D., op. cit., 2017, p.2; Nick Estes, *Our History Is the Future*, London, Verso, 2019, pp.9, 15, 97

22. Nick Estes, ibid., 2019, pp.8, 28; Dee Brown, *Bury My Heart at Wounded Knee: An Indian History of the American West*, London, Barrie & Jenkins, 1970, pp.273, 316, 449

23. Nick Estes, ibid., 2019, pp.77-78

24. Dee Brown, op. cit., 1970, pp.144, 265, 275; Nick Estes, ibid., 2019, p. 78; Derek Wall, op. cit., 2017, p. 31

25. David Graeber & David Wengrow, *The Dawn of Everything: A New History of Humanity*, London, Penguin Books, 2022, pp.20, 55; Dee Brown, ibid., 1970, p.427

26. Robin W. Kimmerer, *Braiding Sweetgrass: Indigenous Wisdom, Scientific*

Knowledge and the Teaching of Plants, London, Penguin Books, 2020, pp.308-09

27. Robin W. Kimmerer, ibid., pp.304-06

28. Derek Wall, op. cit., 2017, p.42; Guy Shrubsole, *The Lost Rainforests of Britain*, London, William Collins, 2023, pp.200-01; *rightsandresources.org/blog/ drc-senate-adopts-new-law-on-the-promotion-and-protection-of-the-rights-of-the-indigenous-pygmy-peoples/* ; *vimeo.com/1010600022/d3515cee82* ; *www. iwgia.org/en/democratic-republic-of-congo/5350-iw-2024-drc.html; www. greengrants.org/2024/08/09/world-indigenous-day-2024/; www.niatero.org/ our-work/grantmaking*

29. *warwick.ac.uk/fac/arts/english/currentstudents/undergraduate/modules/ fulllist/first/en122/lecturelist2017-18/plumwood.pdf*

30. Derek Wall, *Hugo Blanco: A revolutionary for life*, London, Merlin Press/ Resistance Books, 2018, pp.2, 5; *anticapitalistresistance.org/ hugo-blanco-15-11-1934-25-06-2023/* ; *anticapitalistresistance.org/ revolutionary-peasant-leader-hugo-blanco-1934-2023/* ; *climateandcapitalism. com/2020/04/06/angus-interview-in-peruvian-indigenous-newspaper/* For those wishing to find out more about Blanco's ideas, a good starting point is his book *We the Indians: The indigenous peoples of Peru and the struggle for the land*, London, Resistance Books, 2017. See also the 2020 film *Río Profundo* about him: *vimeo.com/ondemand/HugoBlancoFilm*

31. Alan Thornett, *Facing the Apocalypse: Arguments for Ecosocialism*, London, Resistance Books, 2019, pp.70-71

32. Rosa Luxemburg, op. cit., 1971, p.369; Nick Estes, op. cit., 2019, pp.14, 25, 28

33. Robin W. Kimmerer, op.cit., 2020, pp.319-20; Nick Estes, ibid., 2019, pp.122, 124

34. *jksteinberger.medium.com/what-we-are-up-against-2290ba8c4b5c*

35. *jksteinberger.medium.com/what-we-are-up-against-2290ba8c4b5c* ; David Graeber & David Wengrow, op. cit., 2022, pp.4, 9

Chapter 10: A better world *is* possible

1. Naomi Klein, *This Changes Everything: Capitalism vs. the Climate*, London, Allen Lane, 2014, p.vii

2. Michael Löwy, *Ecosocialism: A Radical Alternative to Capitalist Catastrophe*, Chicago, Haymarket Books, 2015, p.12

3. *monthlyreview.org/2004/10/01/capitalism-and-the-environment/*

4. *www.resilience.org/stories/2017-11-17/this-is-blockadia/ www.theguardian. com/environment/2017/oct/11/2017-deadliest-on-record-for-land-defenders-mining-logging; www.globalwitness.org/en/press-releases/more-2100-land-and-environmental-defenders-killed-globally-between-2012-and-2023/*

5. *www.resilience.org/stories/2017-11-17/this-is-blockadia/*; Andreas Malm & Wim Carton, W., *Overshoot: How the World Surrendered to Climate Breakdown*, London, Verso, 2024, pp.49-50 (emphasis in original); *www. coalaction.org.uk/blog_archive/mining-pont-valley-to-end/; reclaimthepower.*

org.uk/fracking/preston-new-road/

6. *www.wealdactiongroup.org.uk/2024/06/horse-hill-supreme-court-judgment/; www.bbc.co.uk/news/articles/cdrlrkz5k2ro; www.greenpeace.org.uk/news/ jackdaw-rosebank-case-update/*

7. *iopscience.iop.org/article/10.1088/1748-9326/abc197/meta*; Andreas Malm & Wim Carton, op. cit., 2024, pp.8, 10-13, 27, 50

8. *media.nature.com/original/magazine-assets/d41586-021-02990-w/19817644*; Andreas Malm & Wim Carton, ibid., 2024, pp.8, 10-13, 27, 50

9. James Hansen, *Storms of My Grandchildren*, London, Bloomsbury, 2011, pp.172, 205, 209; Alan Thornett, *Facing the Apocalypse: Arguments for Ecosocialism*, London, Resistance Books, 2019, pp.106, 108, 109

10. James Hansen, ibid., 2011, p.205; Alan Thornett, ibid., 2019, p.107,108

11. Alan Thornett, ibid., 2019, pp.107-08, 110; James Hansen, ibid., 2011, p.209

12. Alan Thornett, ibid., 2019, p,111; John B. Foster, et al., *The Ecological Rift: Capitalism's War on the Earth*, New York, Monthly Review Press, 2010, pp.116-18, 437

13. John B. Foster, et al., 2010, pp. 437-38

14. *monthlyreview.org/2013/02/01/james-hansen-and-the-climate-change-exit-strategy/* ; Alan Thornett, op. cit., 2019, pp.112-13

15. Alan Thornett, ibid., 2019, p.114; John B. Foster, et al., op. cit., 2010, pp.117-18; *www.peoplespolicyproject.org/project/a-progressive-case-for-a-carbon-tax-and-dividend-scheme/; www.peoplespolicyproject.org/wp-content/ uploads/2018/09/CarbonTax.pdf*

16. Jonathan Neale, *Fight the Fire: Green New Deals and Global Climate Jobs*, London, Resistance Books, 2021,pp.7-8; Naomi Klein, *On Fire: The Burning Case for a Green New Deal*, London, Allen Lane, 2019, p.26

17. *www.ecosocialistdiscussion.com/2019/09/21/green-new-deals-must-push-the-boundaries/*; Naomi Klein, ibid., 2019, p.22

18. *www.ecosocialistdiscussion.com/2019/09/21/green-new-deals-must-push-the-boundaries/; https://anticapitalistresistance.org/starmers-miserable-betrayal-a-gift-to-the-crisis-ridden-tories-and-a-disaster-for-the-planet/*

19. *www.ecosocialistdiscussion.com/2024/02/16/starmers-miserable-betrayal-a-gift-to-the-crisis-ridden-tories-and-a-disaster-for-the-planet/; anticapitalistresistance.org/ review-fight-the-fire-green-new-deals-and-global-climate-jobs/*

20. Derek Wall, *Elinor Ostrom's Rules for Radicals: Cooperative Alternatives Beyond Markets and States*, London, Pluto Press, 2017, p.43

21. Ernest Mandel, *Power and Money: A Marxist Theory of Bureaucracy*, London, Verso, 1992, p.207; Kate Raworth, *Doughnut Economics: Seven Ways to Think Like a 21st-Century Economist*, London, Random House Business Books, 2018, pp.43, 46, 53

22. Giorgos Kallis, et al., *The Case for Degrowth*, Cambridge, Polity, 2020, pp.viii-ix, xi-xii; Jason Hickel, *Less is More: How Degrowth Will Save the World*,

London, Penguin Books, 2020, pp.165, 169-70 (emphasis in the original); Kate Soper, *Post-Growth Living: For an Alternative Hedonism*, London, Verso, 2020, p.1; John B. Foster, et al, op. cit., 2010, p.181

23. *anticapitalistresistance.org/degrowth-%c2%ac-facing-up-to-harsh-reality/*; Giorgos Kallis, et al., ibid., 2020, pp.34-35

24. *climateandcapitalism.com/2023/07/11/the-eco-in-ecosocialism-must-mean-climate-or-we-are-lost/*; *climateandcapitalism.com/2023/09/26/in-defense-of-degrowth-a-reply-to-jonathan-neale/*

25. *www.ecosocialistdiscussion.com/2023/09/26/in-defence-of-degrowth-a-reply-to-jonathan-neale/* ; Ernest Mandel, op. cit., 1992, p.206; Michael Löwy & Olivier Besancenot, *Revolutionary Affinities: Towards a Marxist-Anarchist Solidarity*, Oakland (California), PM Press, 2023, p.125

26. *www.youtube.com/watch?v=LaPge01NQTQ*; *www.youtube.com/watch?v=hPSE6peKCDA*

27. *monthlyreview.org/2015/03/01/a-rational-agriculture-is-incompatible-with-capitalism/*

28. Ian Angus, *Facing the Anthropocene: Fossil Capitalism and the Crisis of the Earth System*, New York, Monthly Review Press, 2016, p.46; Jane Kelly and Phil Ward, 'No Solution Under Capitalism', in Jane Kelly & Sheila Malone, (eds), *Ecosocialism or barbarism*, London, Resistance Books, 2006, p.51

29. Alan Thornett, op. cit., 2019, pp. 173, 182

30. *www.leap.ox.ac.uk/article/reducing-foods-environmental-impacts*; *www.theguardian.com/commentisfree/2018/jun/08/save-planet-meat-dairy-livestock-food-free-range-steak*; *www.nature.com/articles/s43016-023-00795-w*; *greencitizen.com/news/vegan-diets-reduce-environmental-impacts-by-75-oxford-study/*; *www.livekindly.com/global-land-use-beef-vegan/*

31. *viva.org.uk/red-tractor-overview/*; *www.youtube.com/watch?v=oetyotLCGLs*; *www.amazon.co.uk/Hogwood-Jerome-Flynn/dp/B08B8XKGL8/* ; Fred Magdoff & Chris Williams, *Creating an Ecological Society: Toward a Revolutionary Transformation*, New York, Monthly Review Press, 2017, p.255; *www.thebureauinvestigates.com/stories/2017-07-17/intensive-numbers-of-intensive-farming/*; *https://www.bbc.co.uk/news/articles/cy4ldkpz1klo*

32. *www.who.int/news/item/07-11-2017-stop-using-antibiotics-in-healthy-animals-to-prevent-the-spread-of-antibiotic-resistance*; *www.youtube.com/watch?v=bkUU5geAsiE*

33. Alan Thornett, op. cit., 2019, pp.253-54

34. Alan Thornett, ibid., 2019, p.254; *www.ecosocialistdiscussion.com/2024/05/27/green-left-meeting-on-population-april-22nd/*

35. Alan Thornett, ibid., 2019, pp.246, 251-53

36. Alan Thornett, ibid., 2019, p.246; *www.ecosocialistdiscussion.com/2024/05/27/green-left-meeting-on-population-april-22nd/*

37. Ruby Turner, 2019, 'A Better Way', from *Love Was Here*: London, RTR Productions: *www.youtube.com/watch?v=GtDuhFoi9Xo*

38. Wendell Berry, Interview with Bill Moyers, October 7, 2013: *www.alternet. org/2013/10/renowned-farmer-and-writer-wendell-berry-confronting-consequences-runaway-capitalism*; Alan Thornett, op. cit., 2019, p.193; Michael Löwy & Olivier Besancenot, op. cit., 2023, p.110

39. *climateandcapitalism.com/2023/09/26/in-defense-of-degrowth-a-reply-to-jonathan-neale/*

Chapter 11: (*We say we want a*) Revolution

1.   Rosa Luxemburg, *The Accumulation of Capital*, London, Routledge & Kegan Paul, 1963, p.453.

2.   *www.youtube.com/watch?v=BGLGzRXY5Bw; beatlesinterviews.org/db1971.0121.beatles.html*; Michael Löwy, *Ecosocialism: A Radical Alternative to Capitalist Catastrophe*, Chicago, Haymarket Books, 2015, p.39; David Whyte, *Ecocide: Kill the Corporation Before it Kills Us*, Manchester, Manchester University Press, 2020, p.21; Darren McGarvey, *Poverty Safari: Understanding the anger of Britain's Underclass*, London, Picador, 2018, p.109; *www.nature. com/articles/s41586-024-07219-0; www.theguardian.com/environment/2024/apr/17/climate-crisis-average-world-incomes-to-drop-by-nearly-a-fifth-by-2050; www.independent.co.uk/news/world/europe/ap-nature-germany-france-national-oceanic-and-atmospheric-administration-b2530279.html; academic. oup.com/oocc/article/3/1/kgad008/7335889*

3.   Neil Faulkner, et al., *System Crash: an activist guide to making revolution*, London, Resistance Books, 2021, p.6; Enzo Traverso, *The New Faces of Fascism: Populism and the Far Right*, London, Verso, 2019, p186

4.   Michael Löwy & Olivier Besancenot, *Revolutionary Affinities: Towards a Marxist-Anarchist Solidarity*, Oakland (California), PM Press, 2023, p.89; Tony Cliff, *Marxism at the Millennium*, London, Bookmarks, 2000, p.18; *monthlyreview.org/2004/10/01/capitalism-and-the-environment/*

5.   Daniel Tanuro, *Green Capitalism: why it cannot work*, London, Merlin Press/ IIRE, 2013, p.11; Ian Angus, *Facing the Anthropocene: Fossil Capitalism and the Crisis of the Earth System*, New York, Monthly Review Press, 2016, pp.221-22 (emphasis in the original)

6.   Andreas Malm & Wim Carton, *Overshoot: How the World Surrendered to Climate Breakdown*, London, Verso, 2024, p.237 (emphasis in original); Leon Trotsky, *The Transitional Program for Socialist Revolution*, New York, Pathfinder Press, 1977, p.111,

7.   Leon Trotsky, ibid., 1977, pp 114-15 (emphasis in the original); Tony Cliff, op. cit., 2000, p. 86

8.   *www.youtube.com/shorts/vlBKWeh3D2o*

9.   *www.ecosocialistdiscussion.com/2024/09/27/solar-power-the-key-to-the-future-of-the-planet/* ; Michael Löwy, *Ecosocialism: A Radical Alternative to Capitalist Catastrophe*, Chicago, Haymarket Books, 2015, p.23; Andreas Malm & Wim Carton, op. cit., 2024, p.204.

10. Andreas Malm & Wim Carton, ibid., 2024, pp.209, 231-32

11. Andreas Malm & Wim Carton, ibid., 2024, p.p 220-21

12. Michael Löwy, op. cit., 2015, p.1; *www.cnsjournal.org/*; *monthlyreview. org/2004/10/01/capitalism-and-the-environment/*

13. *jksteinberger.medium.com/what-we-are-up-against-2290ba8c4b5c* ; Karl Marx, *The German Ideology*, London, Lawrence & Wishart, 1974, pp. 94-95 (emphasis in original); Leon Trotsky, op. cit., 1977, p.112 (emphasis added);

14. Ulrich Brand & Markus Wissen, *The Imperial Mode of Living: Everyday Life and the Ecological Crisis of Capitalism*, London, Verso, 2021, pp.xiii-xiv; Michael Löwy & Olivier Besancenot, op. cit., pp.45, 112; Michael Löwy, op. cit., 2015, p.13

15. Naomi Klein, *This Changes Everything: Capitalism vs. the Climate*, London, Allen Lane, 2014, p.450; Michael Lebowitz, quoted in Ian Angus, *Facing the Anthropocene: Fossil Capitalism and the Crisis of the Earth System*, New York, Monthly Review Press, 2016, p.212; Fred Magdoff & Chris Williams, *Creating an Ecological Society: Toward a Revolutionary Transformation*, New York, Monthly Review Press, 2017, p.283; Michael Löwy, op. cit., 2015, pp.12-13; John Bellamy Foster, *The Return of Nature: Socialism and Ecology*, New York, Monthly Review Press, 2020, p.7

16. Neil Faulkner, *Mind Fuck: The Mass Psychology of Creeping Fascism*, London, Resistance Books, 2022, pp.117-18

17. *www.youtube.com/watch?v=-ox0uGW3C6k*

18. *Ecosocialist Revolution: A Manifesto*, London, Resistance Books, 2024, p. 24; *www.cacctu.org.uk/climate_jobs_2014*

19. Matt Huber, *Climate Change as Class War: Building Socialism on a Warming Planet*, London, Verso, 2022, p.3; *juststopoil.org/* (emphasis added); *www. theguardian.com/world/2024/apr/11/revolutions-are-coming-who-are-youth-demand-and-what-do-they-want* ; *https://juststopoil.org/*

20. Alan Thornett, *Facing the Apocalypse: Arguments for Ecosocialism*, London, Resistance Books/IIRE, 2019, p.220; *www.globalecosocialistnetwork.net/*

21. Mikaela Loach, *It's Not That Radical: Climate Action to Transform Our World*, London, Dorling Kindersley Ltd., 2024, p.58; Kate Soper, *Post-Growth Living: For an Alternative Hedonism*, London, Verso, 2020, pp.17-.18

22. John Bellamy Foster, *The Ecological Revolution: Making Peace with the Planet*, New York, Monthly Review Press, 2009, p.264; Paul Burkett, *Marxism and Ecological Economics: Toward a Red and Green Political Economy*, Chicago, Haymarket Books, 2009, pp.158-59

23. Neil Faulkner, op. cit., 2022, p.119; Neil Faulkner, et al., *Creeping Fascism: What it is and How to Fight it*, London, Public Reading Rooms, 2019, p.294: Karl Marx, 'Theses on Feuerbach', in *Early Writings*, Harmondsworth, Penguin Books, 1975, p.423

24. Michael Löwy, op. cit., 2015, p.vii; *www.nber.org/papers/w32450; www. theguardian.com/environment/article/2024/may/17/economic-damage-climate-change-report;* Nick Estes, *Our History Is the Future*, London, Verso, 2019, pp.256-7

25. *anticapitalistresistance.org/karl-marx-and-ecology-an-interview-with-michael-lowy/*; Daniel Tanuro, op. cit., 2013, p.16; Rebecca Solnit, *Hope in the Dark: Untold Histories, Wild Possibilities*, Edinburgh, Canongate Books, 2016, p.136; *www.curatedquotes.com/casablanca-quotes/*

26. Paul Burkett, *Marxism and Ecological Economics: Towards A Red and Green Political Economy*, Chicago, Haymarket Books, 2009, pp.54-55, 332; Michael Löwy & Olivier Besancenot, op. cit., 2023, pp.45, 74, 125, 133, 139

27. Michael Löwy & Olivier Besancenot, ibid., 2023, p. 127; Grace Blakeley, *Vulture Capitalism: Corporate Crimes, Backdoor Bailouts and the Death of Freedom*, London, Bloomsbury Publishing, 2024, p.295

Conclusion: Hope and history

1. Rebecca Solnit, *Hope in the Dark: Untold Histories, Wild Possibilities*, Edinburgh, Canongate Books, 2016, p.ix (emphasis added)

2. Ian Angus, *Facing the Anthropocene: Fossil Capitalism and the Crisis of the Earth System*, New York, Monthly Review Press, 2016, p.22 (emphasis in the original); Simon Hannah, *Reclaiming the Future: A Beginner's Guide to Planning the Economy*, London, Pluto Press, 2024, pp.220-21

3. Robin W. Kimmerer, *Braiding Sweetgrass: Indigenous Wisdom, Scientific Knowledge and the Teaching of Plants*, London, Penguin Books, 2020, p.328

4. Stan Cohen, *States of Denial: Knowing about Atrocities and Suffering*, Cambridge, Polity Press, 2001, p.1

5. Robin W. Kimmerer, op. cit., 2020, p.326

6. *socialistregister.com/index.php/srv/article/view/27143/20148*; Ian Angus, op. cit., 2016, p.222; Karl Marx & Friedrich Engels, *The Communist Manifesto*, London, Penguin Books, 1967, p.79 (emphasis added)

7. Ian Angus, ibid., 2016, p.223 (emphasis added); *socialistregister.com/index. php/srv/article/view/27143/20148*

8. *socialistregister.com/index.php/srv/article/view/27143/20148*

9. Allan Todd, *Che Guevara: The Romantic Revolutionary*, Barnsley, Pen & Sword, 2024, pp.xiii, 222; Michael Löwy, 'Ernest Mandel's Revolutionary Humanism'; and Norman Geras, 'Marxists before the Holocaust: Trotsky, Deutscher, Mandel' in Gilbert Achar, (ed.), *The Legacy of Ernest Mandel*, London, Verso, 1999, pp.34, 202

10. Mikaela Loach, *It's Not That Radical: Climate Action to Transform Our World*, London, Dorling Kindersley Ltd., 2024, pp.18, 22-3, 230

11. Rebecca Solnit, op. cit., 2016, pp.1, 5; *fourth.international/en/world-congresses/62*

12. Robin W. Kimmerer, op. cit., 2020, p.376

13. Karl Marx, 'Eighteenth Brumaire of Louis Napoleon', in David Fernbach, (ed.), *Surveys from Exile: Political Writings, Vol. 2*, Harmondsworth, Penguin Books, 1973, pp.236-37

14. Rebecca Solnit, op. cit., 2016, p.xvii (emphasis in the original)

15. Rebecca Solnit, ibid., 2016, pp.xxiii-xxiv, 13

16. Fawzi Ibrahmin, *Capitalism versus Planet Earth: an Irreconcilable Conflict*, London, Muswell Press, 2012, p.245; *www.newstatesman.com/ business/economics/2018/06/if-capitalism-ended-what-would-replace-it*; *newleftreview.org/issues/ii21/articles/fredric-jameson-future-city*; Daniel Singer, *Whose Millennium? Theirs or Ours?*, New York, Monthly Review Press, 1999, p.259

17. Rebecca Solnit, ibid., 2016, p.132

18. *www.youtube.com/watch?v=oo7VlD66ISM*; Shakespeare, W., *The Tragedy of Hamlet, Prince of Demark*, London, Heron Books, 1966, p.512

19. Greta Thunberg, *No One is Too Small to Make a Difference*, London, Penguin Books, 2019, p.4; *getyarn.io/yarn-clip/39e1747c-973b-4828-a2ad-cd50d2472b4e* ; Michael Löwy & Oliver Besancenot, *Revolutionary Affinities: Towards a Marxist-Anarchist Solidarity*, Oakland (California), PM Press, 2023, p.xiv

20. Howard Zinn, *Failure to Quit: Reflections of an Optimistic Historian*, Chicago, Haymarket Books, 2013, p.164 (emphasis added); Ernest Mandel, 'We Must Dream', 1978, in *Ernest Mandel: Hope and Marxism*, (ed. Alex de Jong), London, Resistance Books/ IIRE, 2022, p.122

21. Robin W. Kimmerer, op. cit., 2020, p.328

22. Rebecca Solnit, op. cit., 2016, p.4; Michael Löwy, *Ecosocialism: A Radical Alternative to Capitalist Catastrophe*, Chicago, Haymarket Books, 2015, p.37; Seamus Heaney, *The Cure at Troy: Sophocles' 'Philoctetes'*, London, Faber & Faber, 1990, p.69

BIBLIOGRAPHY

Achar, G., (ed.), *The Legacy of Ernest Mandel*, London, Verso, 1999

Albert, M. J., *Navigating the Polycrisis: Mapping the Futures of Capitalism and the Earth*, Cambridge (Mass.), The MIT Press, 2024

Ali, T., *The Extreme Centre: A Warning*, London, Verso, 2015

Angus, I., (ed.), *The Global Fight for Climate Justice: Anticapitalist Responses to Global Warming and Environmental Destruction*, London, Resistance Books, 2009

Angus, I., *Facing the Anthropocene: Fossil Capitalism and the Crisis of the Earth System*, New York, Monthly Review Press, 2016

Angus, I., *The War Against the Commons: Dispossession and Resistance in the Making of Capitalism*, New York, Monthly Review Press, 2023

Bahro, R., *Socialism and Survival*, London, Heretic Books, 1982

Blakeley, G., *Vulture Capitalism: Corporate Crimes, Backdoor Bailouts and the Death of Freedom*, London, Bloomsbury Publishing, 2024

Brand, U. & Wissen, M., *The Imperial Mode of Living: Everyday Life and the Ecological Crisis of Capitalism*, London, Verso, 2021

Brown, D., *Bury My Heart at Wounded Knee: An Indian History of the American West*, London, Barrie & Jenkins, 1970

Burkett, P., *Marxism and Ecological Economics: Towards A Red and Green Political Economy*, Chicago, Haymarket Books, 2009

Burkett, P., *Marx and Nature: A Red and Green Perspective*, Chicago, Haymarket Books, 2014

Carson, R., *Silent Spring*, London, Penguin Books, 1999

Clare, J., Summerfield, G. (ed.), *Selected Poems*, London, Penguin Books, 1990,

Cohen, S., *States of Denial: Knowing about Atrocities and Suffering*, Cambridge, Polity Press, 2001

Cliff, T., *Marxism at the Millennium*, London, Bookmarks, 2000

Engels, F., *Dialectics of Nature*, Wellred Books, London, 2012

Estes, N., *Our History Is the Future*, London, Verso, 2019

Faulkner, N. & Dathi, S., *Creeping Fascism: Brexit, Trump and the Rise of the Far Right*, London, Public Reading Rooms, 2017

Faulkner, N., et al., *Creeping Fascism: What it is and How to Fight it*, London, Public Reading Rooms, 2019

Faulkner, N., et al., *System Crash: an activist guide to making revolution*, London, Resistance Books, 2021

Faulkner, N., *Mind Fuck: The Mass Psychology of Creeping Fascism*, London, Resistance Books, 2022

Foster, J. B., *The Ecological Revolution: Making Peace with the Planet*, New York, Monthly Review Press, 2009

Foster, J. B., & Burkett, P., *Marx and the Earth: An Anti-Critique*, Chicago, Haymarket Books, 2017

Foster, J. B., *Marx's Ecology: Materialism and Nature*, New York, Monthly Review Press, 2000

Foster, J. B., Clark, B. & York, R., *The Ecological Rift: Capitalism's War on the Earth*, New York, Monthly Review Press, 2010

Foster, J. B., *The Return of Nature: Socialism and Ecology*, New York, Monthly Review Press, 2020

Foster, J. B., & Clark, B., *The Robbery of Nature: Capitalism and the Ecological Rift*, New York, Monthly Review Press, 2020

Graeber, D. & Wengrow, D., *The Dawn of Everything: A New History of Humanity*, London, Penguin Books, 2022

Greger, M., *How to Survive a Pandemic*, London, Bluebird Books, 2020

Hannah, S., *Reclaiming the Future: A Beginner's Guide to Planning the Economy*, London, Pluto Press, 2024

Hansen, J., *Storms of My Grandchildren*, London, Bloomsbury, 2011

Hickel, J., *Less is More: How Degrowth Will Save the World*, London, Penguin Books, 2020

Hornborg, A., *Nature, Society, and Justice in the Anthropocene: Unraveling the Money-Energy-Technology Complex*, Cambridge, CUP, 2019

Huber, M. T., *Climate Change as Class War: Building Socialism on a Warming Planet*, London, Verso, 2022

Ibrahmin, F., *Capitalism versus Planet Earth: an Irreconcilable Conflict*, London, Muswell Press, 2012

Kallis, G., et al., 2020, *The Case for Degrowth*, Cambridge, Polity

Kelly, J. & Malone, S., (eds), *Ecosocialism or barbarism*, London, Resistance Books, 2006

Kimmerer, R. W., *Braiding Sweetgrass: Indigenous Wisdom, Scientific Knowledge and the Teaching of Plants*, London, Penguin Books, 2020

Klein, N., *The Shock Doctrine: The Rise of Disaster Capitalism*, London, Penguin Books, 2008

Klein, N., *This Changes Everything: Capitalism vs. the Climate*, London, Allen Lane, 2014

Klein, N., *On Fire: The Burning Case for a Green New Deal*, London, Allen Lane, 2019

Kolbert, E., *The Sixth Extinction: An Unnatural History*, London, Bloomsbury, 2015

Loach, M., *It's Not That Radical: Climate Action to Transform Our World*, London, Dorling Kindersley Ltd., 2024

Löwy, M., *Ecosocialism: A Radical Alternative to Capitalist Catastrophe*, Chicago, Haymarket Books, 2015

Löwy, M. & Besancenot, O., *Revolutionary Affinities: Towards a Marxist-Anarchist Solidarity*, Oakland (California), PM Press, 2023

Luxemburg, R., *The Accumulation of Capital*, London, Routledge & Kegan Paul, 1971

Lynas, M., *Our Final Warning: Six Degrees of Climate Emergency*, London, 4th. Estate, 2020

Magdoff, F., & Williams, C., *Creating an Ecological Society: Toward a Revolutionary Transformation*, New York, Monthly Review Press, 2017

Malm, A., *The Progress of this Storm: Nature and Society in a Warming World*, London, Verso, 2018

Malm, A., *How to Blow Up a Pipeline*, London, Verso, 2021

Malm, A. & Carton, W., *Overshoot: How the World Surrendered to Climate Breakdown*, London, Verso, 2024

Mandel, E., *Power and Money: A Marxist Theory of Bureaucracy*, London, Verso, 1992

Mandel, E., (ed. De Jong, A.), *Hope and Marxism: Historical and Theoretical Essays*, London, Resistance Books/ IIRE, 2022

Marx, K., *Early Writings*, Harmondsworth, Penguin Books, 1975

Marx, K., *Capital Volume 1*, Harmondsworth, Penguin Books, 1976

Marx, K., *Capital Volume 3*, Harmondsworth, Penguin Books, 1981

Marx, K., *Grundrisse*, Harmondsworth, Penguin Books, 1973

Marx, K. & Engels, F., *The German Ideology*, London, Lawrence & Wishart, 1974

Mason, P., *PostCapitalism: A Guide to Our Future*, London, Allen Lane, 2015

Mason, P., *How to Stop Fascism: History, Ideology, Resistance*, London, Allen Lane, 2021

Mattei, C., *The Capital Order: How Economists Invented Austerity and Paved the Way to Fascism*, London, University of Chicago Press, 2022

McGarvey, D., *Poverty Safari: Understanding the anger of Britain's Underclass*, London, Picador, 2018

Monbiot, G. & Hutchison, P., *The Invisible Doctrine: The Secret History of Neoliberalism*, Allen Lane, 2024

Neale, J., *Fight the Fire: Green New Deals and Global Climate Jobs*, London, Resistance Books, 2021

Packham, C. & Cohen, A., *Earth: Over 4 Billion Years in the Making*, London, William Collins, 2023

Piercy, M., *Woman on the Edge of Time*, London, Del Rey/Penguin, 2019

Pirani, S., *Burning Up: A Global History of Fossil Fuel Consumption*, London, Pluto Press, 2018

Raworth, K., *Doughnut Economics: Seven Ways to Think Like a 21st-Century Economist*, London, Random House Business Books, 2018

Saito, K., *Karl Marx's Ecosocialism: Capital, Nature, and the Unfinished Critique of Political Economy*, New York, Monthly Review Press, 2017

Shrubsole, G., *The Lost Rainforests of Britain*, London, William Collins, 2023

Singer, D., *Whose Millennium? Theirs or Ours?*, New York, Monthly Review Press, 1999

Solnit, R., *Hope in the Dark: Untold Histories, Wild Possibilities*, Edinburgh, Canongate Books, 2016

Soper, K., *Post-Growth Living: For an Alternative Hedonism*, London, Verso, 2020

Stocker, P., *English Uprising: Brexit and the Mainstreaming of the Far Right*, London, Melville House, 2017

Tanuro, D., *Green Capitalism: why it cannot work*, London, Merlin Press/IIRE, 2013

Thornett, A., *Facing the Apocalypse: Arguments for Ecosocialism*, London, Resistance Books/IIRE, 2019

Thunberg, G., *No One is Too Small to Make a Difference*, London, Penguin Books, 2019

Traverso, E., *The New Faces of Fascism: Populism and the Far Right*, London, Verso, 2019

Trotsky, L., *Leon Trotsky on France*, New York, Monad Press, 1979

Trotsky, L., *The Struggle Against Fascism in Germany*, New York, Pathfinder Press, 1971

Trotsky, L., *The Transitional Program for Socialist Revolution*, New York, Pathfinder Press, 1977

Wall, D., *Economics After Capitalism: A Guide to the Ruins, and a Road to the Future*, London, Pluto Press, 2015

Wall, D., *Elinor Ostrom's Rules for Radicals: Cooperative Alternatives Beyond Markets and States*, London, Pluto Press, 2017

Wall, D., *Hugo Blanco: A revolutionary for life*, London, Merlin Press/ Resistance Books, 2018

Wall, D., *Climate Strike: The Practical Politics of the Climate Crisis*, Dagenham, Merlin Press, 2020

Whyte, D., 2020, *Ecocide: Kill the Corporation Before it Kills Us*, Manchester, Manchester University Press

Wendling, M., *Alt-Right: From 4chan to the White House*, London, Pluto Press, 2018

Williams, R., *Resources of Hope: Culture, Democracy, Socialism*, London, Verso, 1989

Wilson, E. O., *Half-Earth: Our Planet's Fight for Life*, New York, Liveright Publishing Corporation, 2016

Zinn, H., *Failure to Quit: Reflections of an Optimistic Historian*, Chicago, Haymarket Books, 2013

Websites

www.ecosocialistdiscussion.com/
climateandcapitalism.com/
monthlyreview.org/
anticapitalistresistance.org/

ABOUT THE PUBLISHERS

RESISTANCE BOOKS is a radical publisher of internationalist, ecosocialist, and feminist books. Resistance Books publishes books in collaboration with the International Institute for Research and Education (iire.org), and the Fourth International (https://fourth.international). For further information, including a full list of titles available and how to order them, go to the Resistance Books website.

info@resistancebooks.org
www.resistancebooks.org/

ALSO FROM RESISTANCE BOOKS

An October in Catalonia: Disruptions and paradoxes of the Catalan independence movement, Josep Maria Antentas.
Published 2025, 300 pages, £18.

Resisting Trumpism, Daniel Tanuro, Paris Wilder, Gilbert Achcar, Simon Hannah, and Echo Fortune.
Published 2024, 126 pages, £8.

Palestine and Marxism, Joseph Daher.
Published 2024, 150 pages, £10.

War, Global Capitalism and Resistance, William I. Robinson.
Published 2024, 200 pages, £15.

Internationalism or Russification: A study in the Soviet nationalities problem, Ivan Dzyuba.
Published 2024, pages, £17.

Israel's War on Gaza, Gilbert Achcar.
Published 2023, 98 pages, £6.

Hope and Marxism: Historical and Theoretical Essays,
Ernest Mandel.
Published 2023, 306 pages, £15.

Introduction to Marxist Theory, Selected writings, Ernest Mandel.
Published 2021, 306 pages, RRP£15.

Making Sense of Russia's Invasion of Ukraine, Paul Le Blanc.
Published 2024, 95 pages, £6.

Capitalist China and Socialist Revolution, Simon Hannah.
Published 2022, 78 pages, £6.

Radical Psychoanalysis and Anti-capitalist Action, Ian Parker.
Published 2022, 82 pages, £6.

Mind Fuck: The Mass Psychology of Creeping Fascism, Neil Faulkner.
Published 2022, 136 pages, £6.